Samuel Findley

Rambles among the Insects

Samuel Findley

Rambles among the Insects

ISBN/EAN: 9783743327177

Manufactured in Europe, USA, Canada, Australia, Japa

Cover: Foto ©berggeist007 / pixelio.de

Manufactured and distributed by brebook publishing software
(www.brebook.com)

Samuel Findley

Rambles among the Insects

RAMBLES

AMONG THE INSECTS.

BY THE

Rev. SAMUEL FINDLEY, D. D.,

CORRESPONDING MEMBER OF THE AMERICAN ENTOMOLOGICAL SOCIETY.

"O Lord, how manifold are thy works! in wisdom hast thou made them all."—Ps. civ. 24.

PHILADELPHIA:

PRESBYTERIAN BOARD OF PUBLICATION,

No. 1334 Chestnut Street.

TO HIS

DEAR WIFE

THIS VOLUME

IS MOST AFFECTIONATELY DEDICATED

BY

THE AUTHOR.

WESTCOTT & THOMSON,
Stereotypers and Electrotypers, Philada.

INTRODUCTION.

If my young friends will peruse this book with half the pleasure I enjoyed whilst gathering the facts which it relates, I shall be abundantly repaid for the labor expended in writing it. The air and fields are so full of beautiful and sprightly things of life that I shall not fail to confer a favor upon intelligent and inquisitive boys and girls if I succeed in inspiring them with an interest in the study of insect-life. The objects embraced in this study are within the reach of every one. They intrude themselves into our presence uncalled, and at times are most unwelcome. For this reason they are despised, and their structure, instincts and domestic habits are unstudied. But, like all the works of God, they display his wisdom and goodness, and are, on this account, worthy of the profoundest study.

I have endeavored to present the history of insects, embracing their early life, their structure, instinct, intelligence, cunning, their architectural skill, habits and customs at home, care of their young, modes of capturing their prey and their wonderful transformations, in a simple and attractive style, that my young friends may be induced to enter upon the study of insect-life, and to prosecute it till their familiarity with the subject shall impart to them all the enthusiasm of young naturalists.

In addition to these facts, so interesting in themselves, I have directed special attention to the testimony they bear to the goodness and wisdom of God in adapting the structure and instinct of insects to the preservation and happiness of the individuals. I have also sought to point out the important lessons of industry, perseverance, endurance, economy and foresight which their domestic life and habits teach. With what success I have accomplished my undertaking the careful and interested reader must decide.

I embrace this occasion to acknowledge my great indebtedness to MR. C. A. BLAKE, Corresponding Secretary of the American Entomological Society, for very important assistance rendered me in the preparation of this work. Before the manuscript was put into the hands of the Presbyterian Board of Publication it was submitted to his critical examination and review, and his criticisms and suggestions were incorporated in the book. He has also read the revised proof-sheets, and has thus given me the benefit of his extensive and accurate knowledge in making it reliable in its classification and statement of scientific facts.

I am also much indebted to the Board of Publication for the neat and elegant style in which it has presented this book to the public.

THE AUTHOR.

CONTENTS.

CHAPTER X.

CHAPTER XI.

CHAPTER XII.

CHAPTER XIII.

CHAPTER XIV.

8 *CONTENTS.*

CHAPTER X.

CHAPTER XI.

CHAPTER XII.

CHAPTER XIII.

CHAPTER XIV.

Rambles among Insects.

CHAPTER I.

ENTOMOLOGY, WHAT?—INSECTS, WHAT?—EGGS OF
INSECTS—LACE-WINGED FLY—SAW-FLY—DRAGON-
FLY.

ONE very pleasant evening in the month of April
Uncle Samuel was sitting on the piazza admiring
the beauty of the vegetation, which was just putting
on its spring attire, and listening to the merry songs
of the birds, that had so lately returned to their
former haunts, and were industriously engaged in
building cozy little homes for themselves and their
families. But Uncle Samuel was not to remain undisturbed, for just then his sister's children, Charlie,
Henry, James and Mary, coming home from school,
broke in upon his meditations by their usual inquisitiveness.

"I found a word in my lesson to-day, uncle,"
said James, "that I didn't understand. Do you
think it right to put such big words in books made
for children as young as I am? Such words just
puzzle boys' brains, if they have any. I don't

11

think it fair. I know I just blunder on them, that's all, and I don't know what they mean when I do pronounce them. Our reading lesson was about some *thing* or some *place* called *Entomology*. I don't know whether it is a country or not. If it is, it would be a good place to go to catch all kinds of insects, for there are a great many found there— some very beautiful, and others just as ugly and bad as they can be. Now do tell us what *Entomology* is."

Now, Charlie had studied a little Latin and Greek, and had learned something about the derivation of words, and how the meaning of compound words is taken from the parts of which they were composed. So, as soon as James was through, he said,

"I think I can answer that question myself. Entomology comes from two Greek words,* and is the name of the science that treats of insects. It tells us about their structure and their habits and instincts, just as geography tells us about the earth and its rivers and oceans and mountains and plains and countries.—Am I right, uncle?"

"Yes, you have given us its meaning correctly, and I am glad to see that you make a practical use of your knowledge of language. And as to putting 'big' words in books for children, of which James complains, it is all right when they are better than any other words that can be used. When you know

* *Entoma*, insects, and *logos*, discourse.

their meaning, 'big' words are as easily understood as 'little' words, and often more expressive. Do you understand what entomology is now, James?"

"I think I do. It isn't a place where any person lives and where insects are found, but it is a study like arithmetic and geography. It won't puzzle my brains any more."

"Since you have mentioned entomology," said Uncle Samuel, "I remember my promise that when I would come to see you I would take a ramble with you occasionally among the insects. You are surrounded by beautiful woods and valleys and meadows, where insects like to live, so that we may visit their homes and learn how they are built. I can tell you a great many interesting stories about how they keep house and about their music—for there are several bands of musicians among the insects—and about the way they walk and fly and leap, and about their wisdom and their cunning tricks."

"Oh, do tell us!" said Mary. "I like stories so much. But what *are* insects? Do you mean the little ugly bugs I hate so much? I am sure they won't make nice stories. Mother says bugs were never made for any good, but only to vex us and make us angry."

"You ought not to hate anything, my dear Mary, that a kind and good God has made. He made all things very good. It is sin that has made anything bad. When I speak of insects, I include bugs, for bugs are part of the great household of insects.

But ugly and bad as many of them seem to you, if you only knew more about them you would love and admire them. There are some very ugly and bad boys and girls, who seem to live only to torment and vex their schoolmates; but you don't hate all boys and girls on that account.

"But I must answer your question about insects. *An insect is a small animal whose body is made up of several parts, like rings jointed together, so as to move on each other.* A caterpillar is an insect, because it is made up of several ring-like divisions which it draws into each other when it is in motion. A wasp is an insect. Its head is joined to its body by a fine thread, and its body is joined to its abdomen by a very slender cord. Bees, ants, butterflies, beetles, crickets, dragon-flies and mosquitoes are all insects. Among such interesting little creatures as these we shall have our ramble, picking one up here and there, examining it, learning all we can about its structure and its history, and what kind of habits it has."

"Why are flies and bees and ants and all the other things you named called *insects?* Wouldn't some other name do just as well?" asked fidgety little Miss Mary, who had been restless for some time, waiting for a stopping-place in Uncle Samuel's long talk.

"That's right, little inquisitive; don't let me go on too fast for you. There is a reason why the word insect is used as the name of these little

animals rather than any other word, and Mary
ought to know it."

"I think I know the reason," said Charlie. Now,
Charlie took great delight in finding out the reason
of things, and was specially interested in the study
of words, and was always ready to give his opinion.
"It is derived from a Latin word * which means
'cut into,' and I think it describes flies and bees
and wasps and such creatures better than any other
name, for some of them look as if they were 'cut
into' in more places than one. Look at this fly,
Mary; its head is almost separated from its body.
Does it not look as if somebody had almost cut it
in two?"

"Oh, I see now," said Mary; "they just make believe somebody has cut them in two, and call themselves insects. I understand."

"I am glad Charlie has so well convinced you
that these familiar friends whose homes we propose
to visit have been wisely named. You will find it
capital fun and healthful exercise to chase butterflies and hunt beetles under stones, and to watch
caterpillars when they are eating and spinning, and
wasps building their houses. I will go with you to
the woods sometimes, and we shall have many
glorious rambles together."

"I remember," said Henry, "how James and I
chased butterflies last summer in our meadow. One
day we saw a beautiful one with black wings

* *Insectum,* from *inseco,* to cut into.

marked around the edges with yellow spots, and
a tail divided like a swallow's. When I caught
it, James cried out, 'Oh how splendid! Isn't it
a beauty? The dear little thing! How I wish I
could pet it and have it all for my own!' But it
was a beauty, though! Well, we put it under a glass,
and it fluttered about and knocked almost all the
paint off its wings and broke pieces out of them.
I was so sorry. One day mother took the glass
away, but the poor thing did not live long to enjoy
its freedom."

"I heard you say once that you hated caterpillars, they were so ugly," said Uncle Samuel. "Now,
that very beautiful butterfly was in its early
life a caterpillar, and it was just as splendidly
adorned when it crawled as when it flew. When
it was a caterpillar, its body was a brilliant green,
having a yellow stripe on each side and a row of
blue dots, while its under side and its feet were red-
dish. It feeds on the leaves of the sassafras, where
you may happen to find it in the summer. This
beautiful and tasteful caterpillar will excite your
admiration when you study it as much as you did
the butterfly you captured.

"I am glad that you observed the markings of
your imprisoned butterfly as closely as you did. I
want you to go through the world with your eyes
open, and your minds awake and ready to learn.
There are beautiful objects all around you if you
would but look at them. There are lessons to learn

from the tiniest insect if you would only read the book that nature spreads wide open before you. And there is no department in nature you can study more easily than the insect-world. It is peopled with immense multitudes of interesting objects, which you can readily collect and examine, and by close observation you can become familiar with their various instincts and habits.

"One of the wonderful things about insects is that they follow different trades and employments as men do. But they are not all industrious. Some are idle, and, like idle men, have but little of their own to live on. A man who was well acquainted with insect-life describes their trades and occupations in this way : ' Here we see the industrious laborer busy at his work, there the lazy, lounging beggar ; here, up on the leafy boughs or before the gates of their subterranean abodes, myriads of musicians are playing their fiddles, and there the skillful architect is building his wonderful dwelling; while above, in the blue sky, flutters a high nobility clad in gold, silver, purple and silk, fed on the nectar of flowers, and on the earth below are lurking troublesome drones and disgusting parasites.' "

"What do you mean by *subterranean abodes ?*" asked little Mary.

"Sub-terra*nean* is the word, and it means *under the ground*, and *abodes* means *homes*. ' Subterranean abodes ' are the homes that some insects make for themselves under the ground. Men have such homes out West. They call them 'dug-outs.' Insects that live in 'dug-outs' sometimes open their doors and sit down before them and fill the air with their music. Their music is not very attractive to us, but it pleases the taste of the little creatures who make it."

"But what's the use of learning anything about insects," continued Mary, "when mother says they are good for nothing but pests?"

"Your mother means the mischievous insects which destroy our gardens and eat up what we want for our own use, or annoy us with their unwelcome music, or jag us with their poisonous stings. But all insects are not pests. There are a great many good insects. Some are used by us for coloring-matter and medicine and food, and some are kind enough to spin fine silk for thread and clothing; some spend all their days taking away dead animals and other filth, so as to keep the air we breathe pure and save us from sickness and death. Now, we ought not to class all these friendly insects with those that are doing us all the harm they can. We should know our insect friends as well as our insect enemies. There is a very good reason, therefore, why we should learn all we can about insects, so that we may keep our enemies from doing us injury, and give our friends a chance to do us all the good they can. There is another thing we must not lose sight of in our study of insects, and that is the wisdom and goodness of God in giving foresight and skill to them, so that they not only take good care of themselves, but know how to provide for their offspring. We must never forget that all the insects are praising God every day, and we ought to learn from them to love and praise him daily."

At the invitation of their uncle the children went into his studio, where he had his fine microscope, with which he was ready to show to his interested young friends enlarged views of the minute parts of insects, and thus more fully exhibit to them the wonderful skill and wisdom of their Creator. He placed a few insect eggs under the microscope and called his young pupils to examine them.

"How beautiful!" exclaimed Henry. "They are carved as though some skillful artist had made them. Who would have thought that such splendid eggs could come from insects?"

"You see," said Uncle Samuel, "how the good God takes delight in making his works, even the most minute, very beautiful. It is just as easy for him to adorn an insect's egg with carved work and with different colors, such as white, orange, blue, red and green, as to make them smooth and white, or black only. I will show you a variety of eggs, each one possessing a shape and color and carving peculiar to itself. But here is one that has a special arrangement for the convenience of the little worm when it is ready to leave its birthplace and go out into the world to make its own living. It has a lid, which the worm has just opened on its way out of the egg. Now, this lid is a contrivance showing design and proving the wisdom and skill of the Creator in the structure of the egg. It has a hinge by which it is fastened to one side of the opening, and while the little worm is growing within the egg the lid is secured to the other side of the opening by strong fastenings. When the worm is ready to escape, it knows just how to unfasten the door that confines it in its prison. The lids of some insects' eggs are so fastened to all sides of the opening that the worm pushes them open by force; others are held down by a spring, which the insect touches just in the right way and the right time."

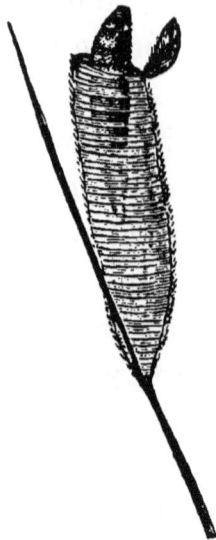

Fig. 1.—An Egg with a Lid.

"Are all insects' eggs provided with such a convenient door for the escape of the little worm?" asked Charlie.

"Not all. Very many insects are under the necessity of gnawing their way out of the egg very much as the little chick pecks its way out of its egg. Insects that pass the first period of their life in water—like the young of the gnat and mosquito—always make their escape from the lower end of the egg directly into the water. This is a wise instinct; for if they should make a mistake and crawl out on the ground, they would die, because they cannot breathe out of water."

"Does their mother take care of the little baby-insects and feed them?" asked Mary.

"The mothers of most insects," answered Uncle Samuel, "never see their offspring, and would not know them if they did, they are so unlike their parents in shape and habits of life. So they do not feed their little baby-insects; but some insect-mothers lay up enough good, rich food for their young just where they can get it as soon as they leave the egg, and others lay their eggs on the trees and plants, on the leaves of which their young worms feed. The butterfly has no relish for the coarse food on which its caterpillar lives and grows; yet when she is ready to lay her eggs, as though she remembered what she ate when a young crawling worm, she carefully selects the very kind of tree or plant on which she once fed, and there deposits her eggs. So when the young leave the egg they find themselves surrounded with the very kind of food they need. Is not this very kind and thoughtful in the butterfly?

"But this is not all that the mother does for her offspring. She provides against any injury that may threaten her eggs by covering them with a substance like glue, which fastens them to the limb and protects them from being washed off by hard rains. Some insect-mothers pluck from their own bodies all the soft down, and with it cover their eggs, so as to protect them from freezing during the cold

winter. Then, having done all they can for their
offspring, they lie down and die.

"There are certain insects that are very fond of
the eggs of certain other insects, and eat them whenever they can find them. The lace-winged fly,
called *Chrysopa perla*, seems to be aware of this
fact, and she has been taught by her Creator a very
cunning way of placing her eggs out of the reach

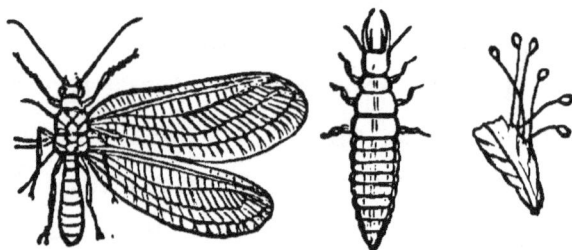

Fig. 2.—THE *Chrysopa perla*, OR LACE-WINGED FLY OF EUROPE.
The perfect insect, the larva, the eggs.

of such egg-loving insects. The cut represents the
eggs, larva and adult of this fly. The wings of
one side only are seen.

"The young of the perla live on plant-lice, so
the thoughtful mother selects a plant infested by
lice for the roving-place of her offspring, that they
may find plenty of food near them as soon as they
enter the world. Her next care is to protect her
eggs from their enemies. To effect this she fastens
a long slender hair to a leaf, at the end of which
she places a small orange-colored egg. A group
of eggs is thus fixed to a single leaf, so as to seem
to be a small tuft of moss in blossom, while the
eggs are far out of the reach of their foes. As soon

as the young grub appears it crawls along the hair to the leaf and commences to eat the plant-lice; and being provided with a pair of large, curved, sharp teeth, moving sidewise, each pierced with a hole, through which it sucks the juices of its victims, it makes great havoc among the ugly things.

"Some insects have an instrument like a saw, with which they cut a slit in the bark of a young shoot, in which they lay their eggs, covering them immediately with a greenish fluid taken from their mouth, and which hardens on exposure to the air. Mr. R. H. Lewis tells a wonderful story about a saw-fly which he observed in Australia. The mother deposits her eggs in a slit in a leaf, and then sits on the leaf till the grubs—the young saw-flies—are hatched. After they leave the egg she follows them, covering them with her legs just as a hen covers her chickens, and protecting them from their enemies with kind motherly tenderness and perseverance.

"The little young dragon-fly, or snake-feeder, lives under water, while its mother dwells in the air and cannot live a minute in the water. How do you think she is able to provide for her offspring? I will tell you. Her tail is very long. When she is ready to lay her eggs, she flies to the side of some pond and lights on a water-plant. Here she makes her observations. If it is a suitable plant for her purpose, she climbs down the stem till she reaches the surface of the water. If you were near and could understand her language, you would be very apt to hear her saying to herself something like this:

"'My little children, when they come out of the egg, cannot live in this bright, clear atmosphere. They must live and grow at the bottom of the pond, just as I did in the days of my grubhood. I will therefore lay my eggs on this plant under the surface of the water, so that when they hatch out they will be just where they will spend the days and nights of their early life.' And when she closes her little speech, she puts her long body in the water and lays her eggs on the plant, some distance below the surface."

"What wonderful stories you tell, uncle, about the cunning and forethought of insects!" said Henry. "I think I shall become very fond of the study of insect-life. It seems to me they must have a real good time among themselves. And their mothers are so kind, providing food and everything nice for their children. But then they don't live to see their children, poor things! I should think the young insects would wonder where they came from and how they got there, with no mother to tell them."

"Isn't it strange," said James, "that they all know just what to do as soon as they get out of the egg, and haven't to be nursed as we are when we are young? I think they must be very wise indeed."

"I 'spect God tells them," said Mary. "He knows."

"Very well said, little thoughtful," said Uncle Samuel with a smile of approval. "That's about all we can know just now about how these little insects come to have so much knowledge. Now, since you are all so much interested in the wonderful facts I have told you, I think we can promise ourselves a rich time during my long stay with you. I have brought with me my fine collection of home and foreign insects, and a number of interesting books on insects, and my microscope and butterfly-nets, and we will have many talks and rambles together, which I know you will enjoy. And while you will be finding out curious facts about insects, you will also be learning more than you have ever known before concerning the wisdom and goodness of God in teaching insects how to take care of themselves, to build their own houses, and to provide for their own offspring."

CHAPTER II.

"WELL, I can't understand it, anyhow. Uncle says that the little baby-butterfly is a worm —a caterpillar—and not a butterfly at all. Now, how can that be, when it is a *butterfly* that lays the egg, and not a caterpillar? I'd like to know that. I think an ever-so-little butterfly ought to come out of a butterfly's egg, just as a little chick comes from a hen's egg. Uncle would not tell what was not true, I know, but I can't just see how it can be as he says."

Mary was thus arguing with her brothers against what had been told them at the last conversation about the young of the butterfly being so very different in appearance from its mother that she would not recognize it as a member of her family if she saw it, when Uncle Samuel appeared and took his seat

in the cozy study-chair his sister had kindly provided for him. It was not long before he fully
understood the nature of the controversy and the
position Mary had taken; and drawing her to his
side and putting his right arm around her affectionately, he said :

"I am so *very* much pleased with my little niece
because she is not willing right away to believe
everything just because, and *only* because, her uncle
says it is so. While his word ought to be good evidence, I am glad she wants to know how he knows
that the strange facts he has told about insects are
true. It does seem as if a little butterfly should be
hatched out of a butterfly's egg.

"I heard a school-teacher once tell a house full
of people that he found a butterfly's egg once on a
hop-vine, and that he put it away, and after a long
time a beautiful butterfly came out of it. Now, if
a teacher of little girls would tell such a story as
that, I do not wonder that my little niece would
believe it to be true. But it wasn't a butterfly's *egg*
that he found—it was a butterfly's *chrysalis;* and
that was the mistake he made.

"The way we know that a caterpillar is hatched
from a butterfly's egg is by seeing the little worm
coming out of the egg, and watching it growing
bigger and bigger until it quits being a caterpillar,
and, changing its form altogether, it becomes a
beautiful butterfly, sailing on its broad, light wings
like

'A winged flower or a flying gem.'

I did not tell you any 'big story' just because you were young and might be deceived, but a strange, yet a true, story."

"Tell us, uncle, about the caterpillar. Why do people call the little butterfly a *caterpillar?* I think it ought to be called its *right* name if it isn't like its mother." So reasoned Mary.

"Well, my young philosopher, you seem to be very critical. If a young hen is called a chick, and a little child is called a babe, and not a man, you ought to expect that the young of the butterfly would have some different name given it. Its name is derived from its character. It is a great destroyer of the leaves of garden-plants and trees; hence the English people gave a name to it made up of two words borrowed from the French, and which mean, when put together, *garden-robber* or *plant-destroyer.* The French people call it by a name that means *evil-doer.* So you see its bad character has given it a bad name. You should all remember this, for, like the caterpillar, boys and girls make a name for themselves by the kind of character they have. Caterpillars are great eaters. They often consume in twenty-four hours double their own weight in food. Is it any wonder they grow rapidly? Their form is well suited to their mode of life, and they have some singular habits which make them sometimes more interesting objects of study than when, gifted with their large and beautiful wings,

' They through the blue air wander.'

"The body of a caterpillar is very soft and pliable, made up usually of twelve segments or rings. These rings fold on each other when the caterpillar walks. The head is very hard, and is provided with strong and powerful jaws, which cut the leaves on which it feeds very much as a pair of scissors would do. These jaws are called its *mandibles.* I want you to learn and remember the names by which the different parts of an insect are known by persons who write about them, so that you can read and understand the books that are written on this subject. Charlie will find use for all that he knows of Latin and Greek in the study of insects, because the terms used in this study are derived chiefly from these languages."

"Can a caterpillar see?" asked Mary. "I never saw any eyes in its head."

"By close inspection you will find in a circle on each side of a caterpillar's head six small black spots. These spots are supposed to be its eyes. We know it is not blind, for it searches for its food and knows just where to go for it. There are some grubs that are hatched in close, dark rooms, and have their food put in these rooms and lying all around them, and who do not need any eyes. These are blind till they get their wings, and then the good God gives them large and beautiful eyes, for they could not enjoy life without them."

"I have often wondered how a caterpillar

breathes," said Henry. "I am sure it has no nose, like a horse or a dog."

"It has that which answers the same purpose, if it is not the same in shape and position. Take up the first caterpillar you find and look carefully along its side a little below the middle, and you will discover a row of *spiracles*—that is, breathing-holes—on each side of the caterpillar. These spiracles are found on nine of the segments, or rings composing its body, two on each segment, making eighteen in all. Each spiracle is surrounded with a colored ring, which adds to the beauty of the caterpillar. So, despised as a caterpillar is, it is provided with more noses than you are. Whether it has the sense of smell or not is not so easy to decide.

"Caterpillars are not all colored or marked alike. Their dress depends upon their species. Some are smooth and almost as transparent as glass; others are covered with hair, thickly set or in tufts; while some have horn-like spines, which make them look very frightful indeed. Some very homely caterpillars grow up into very handsome butterflies, so that you cannot conjecture from the appearance of the caterpillar what will be the size and color of the butterfly unless you are acquainted with its entire history.

"The young of the beetle are called *grubs*, and the general name of the young of all insects is *larvæ*. This word is the plural of *larva*, a Latin word that means *mask*. This name is used because in the caterpillar and grub the perfect insect is supposed to be covered up as by a mask. Indeed, this has been proved to be the fact by an eminent naturalist who dissected a caterpillar, and found enclosed in its body all the parts of the perfect butterfly just waiting its time."

"Does the caterpillar grow up into a butterfly just as a boy grows to be a man?" asked James. "If so, I think we would see its wings just beginning to grow, and its long legs, just as the beard begins to grow on a boy's face. I can't see how it changes its form so much."

"You remember," answered Uncle Samuel, "that strangely-formed thing you once found hanging on the hop-vine, and how you admired it because of its beautiful gold markings? That was once a caterpillar, and lived on the leaves of the hop. When it had grown to its full length and its caterpillar days were all numbered, it fastened itself to a branch of the vine, where it gradually changed itself into the form which attracted your attention. Now it is neither a caterpillar nor a butterfly, but the butterfly is shut up in its golden-marked case, and is only waiting spring-time to break away from its prison and fly away through the air, a more handsome and joyous insect than it ever was.

"This is the third stage in the life of the butterfly, and it is now called *chrysalis*, which means a *golden sheath*, from a Greek word that means *gold*. It takes its name from the golden spots which are on the outside of the sheath. It is also called a *pupa*, which is the Latin word for *an infant*, because the butterfly is wrapped up in its covering as an infant is wrapped in its first clothing, which covering it casts off in the proper time.

"Here is a picture (Fig. 3) of the caterpillar, and

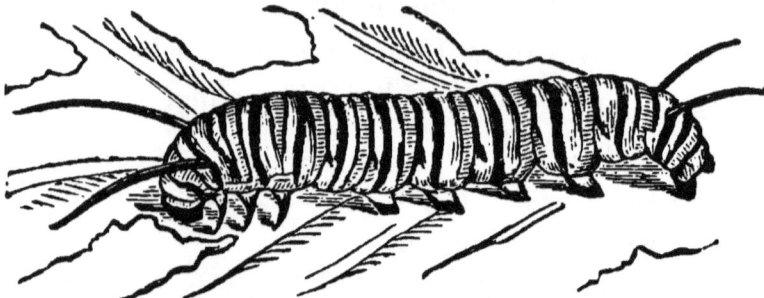

Fig. 3.—CATERPILLAR OF THE ARCHIPPUS BUTTERFLY.

below (Fig. 4) of the pupa, of the Archippus butterfly, a very common tawny-orange butterfly. The caterpillar has black, yellow and white bands on its body, and a pair of threadlike black horns on the top of the second segment, and a shorter pair on the

eleventh segment. The chrysalis is about an inch long, and is usually suspended to the under side of a leaf. Its color is green, and it is ornamented with black and gold spots.

"Some caterpillars of moths make for themselves houses by drawing leaves together by silken threads; and inside of these leaf-houses they change into the pupa state. These houses are called *cocoons*. Some caterpillars that

Fig. 4.—ITS CHRYSALIS, OR PUPA.

are covered all over with hair take the hair off their body and make it up into a cocoon, where the chrysalis is kept safe and snug till the butterfly comes out of it. Here is a drawing of the Isabella tiger-moth (*Arctia Isabella*) in its three stages. You have seen this hairy caterpillar oftentimes. Its

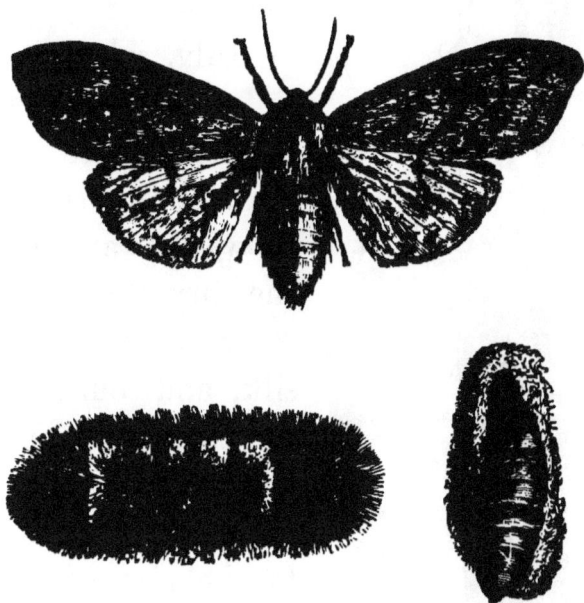

Fig. 5.—ISABELLA TIGER-MOTH (*Arctia Isabella*) IN ITS THREE STAGES.

hairs are like bristles, and on the ends of the caterpillar they are black; the middle hairs are tan-red. When it is caught up in the hand, it rolls itself up in a ball just like a hedgehog."

"I know that caterpillar," said Henry. "I have taken up many a one, and they would roll right out of my hand. I have found them under boards in the winter—dead, I suppose, for they would not move, and it was cold enough to freeze them."

3

"They remain in a half-torpid state during the winter," said Uncle Samuel, "and in April or May they make an egg-shaped cocoon, composed chiefly of their own hairs, where the insect rests till June or July, when it appears as a moth. The cocoon is represented in the drawing as cut open to show the appearance of the chrysalis.

"Some moths spin fine silk, and build a cocoon for themselves out of this silk. These cocoons are large or small according to the size of the moth. This picture (Fig. 6) represents the cocoon of one of the largest of moths. It belongs to the Cecropia moth, and is made for the safe-keeping of the pupa, which is carefully placed within it. When weaving this strange covering, the long green worm seems to know that out of it, in a

Fig. 6.—COCOON OF THE CECROPIA MOTH.

few months, it will want to go forth, released from
confinement; so it is careful to weave the upper end
of the cocoon very loosely, so that it may be able to
push the threads aside very easily when it is ready
to escape. Inside of this curiously-woven house the
caterpillar changes itself into a light-brown chrysalis. When it is ready to enter into the world again,

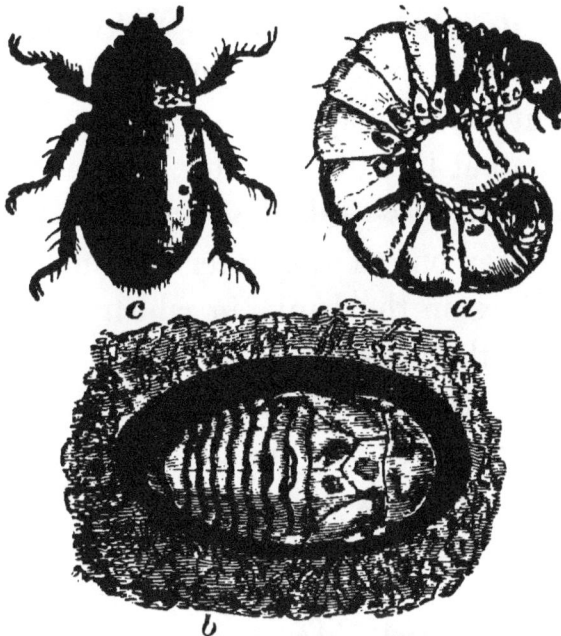

Fig. 7.—THE SPOTTED PELIDNOTA (*Pelidnota punctata,* LINNÆUS).

a fluid flows from its mouth that dissolves the gum
that holds the strands of the cocoon together, and
it soon pushes itself out into the light.

"The grubs of beetles are not as handsome as
caterpillars. As they do their work chiefly in the
dark, they are not ornamented with a variety of
colors, and they are usually clumsy in shape. I·
have a drawing of a beetle you have often seen

making its breakfast on the leaves of the grape-vine. (Fig. 7.) It is here represented as a grub (*a*), pupa (*b*), and perfect insect (*c*). The pupa is often found in rotten stumps and roots of the pear tree. In preparing for its pupa state, the grub forms a rude and not very substantial cocoon out of its own excrement, mixed with the rotten wood that surrounds it. This insect is supposed to live as a grub upward of three years, and in the pupa state about eight or ten days.

"Larvæ that live in the water, such as the young of the *diving-beetles*—called *water-tigers* from their ferocity—creep out of the water and make a house for themselves in the bank, in which they undergo the change that fits them for their higher life. The family name of these beetles is *Dytiscidæ*. If Charlie will look into his Greek lexicon, he will find the word *dutikos*, meaning *fond of diving*. From this Greek word this family of beetles derive their name. The larva of the dragon-fly undergoes but little change when it becomes a pupa, and is still active. When about to make its final change, the pupa quits the water; and clinging to a branch of some water-plant, there it remains till it splits open in the back, and comes forth adorned with those narrow and beautiful wings which carry it so proudly and swiftly through the air.

"The tobacco-worm, with which you are familiar, works its way into the ground, and there is changed into a brown chrysalis, having its long proboscis in a slender sheath that resembles the handle of a pitcher. There it remains all winter; and when the sun warms the earth and brings out the flowers in the spring, it leaves its grave-clothes in the ground and goes forth, a large and beautiful moth, to drink the sweets of flowers and to plague the tiller of the soil."

Now, Mary was a very attentive listener to these strange stories; and though young, she was quite a philosopher in her way. Being sympathetic also, as all little girls are, she began to feel for the welfare and comfort of the poor worm that had shut itself up in its chrysalis and wrapped a shroud of silk about it, or buried itself in the ground to await, through cold and rain and snow, the coming of spring-time for its release. So, looking up into her uncle's face, as if almost doubting the truth of what he was telling them, she said,

" But look-ee here, uncle : you must not forget that the caterpillar can't live without breathing; and how can it breathe if it is shut up in such a close prison, where it doesn't move for so long a time? I'd like to know that."

" Yes, and *I'd* like to know *how* it eats, and *what* it eats, when its mouth is closed up and it is buried in the ground or covered up in its cocoon."

These words came from James, accompanied with a significant flourish of the head, as if he was confident that he and Mary had completely demolished all the stories their uncle had told about caterpillars. Henry was more thoughtful and inclined to take the facts for granted, and to seek for a theory that would reasonably account for them. So, turning to James, he said,

"Don't put on such airs, James, as if you knew everything. If the camel of the desert can live several days without drinking, and we can sleep during a long night without food, the caterpillar might sleep all the winter long without getting hungry. That may be its nature, you know. But I don't exactly see how the pupa can breathe."

Charlie, recollecting that he had dug up in the garden a chrysalis of the potato-worm, thought that this would be a good time to examine it, and perhaps, with it, the problem suggested by Mary might be solved. So he proposed to bring it into the study, that they might search for breathing-holes. This was just what the little naturalists wanted, and they waited with some impatience the few moments during which Charlie, with hurried step, went to the summer-house, where he had laid away this interesting specimen of natural history, and returned.

"Ah! Here they are," said Charlie. "Spiracles like those in the caterpillar, though not colored as they are. It breathes through openings in the side of the abdomen, which moves when I touch it. I suppose Henry is right about its not eating. It is passing into a new state, and does not need to eat."

Uncle Samuel listened with great interest to this

discussion, because it proved that his young pupils were disposed to think and inquire for themselves—characteristics very necessary to the successful study of any subject. It was now his time to speak, and

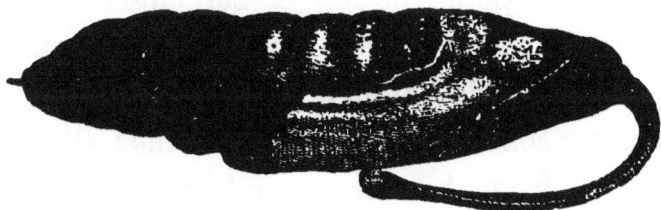

Fig. 8.—PUPA OF POTATO-WORM.

a pause in the conversation indicated a conviction on the part of the young naturalists that they had exhausted their store of ideas.

"I have not forgotten, Mary, that the young butterfly must breathe in order to live even when imprisoned in its pupa-case. Charlie has shown you that the good God has not forgotten this, either, and that he has made provision for its continued breathing. He never neglects the wants of any of his creatures, however small they may be; and when it is necessary for them to undergo great changes before they reach their perfect state, he takes care that these changes do not take away their life. So he supplies the pupa with air as freely as he does the caterpillar, and in most cases by means of the same kind of arrangement. As to the pupa's living without eating, it is one of the strongest proofs of the wisdom and goodness of God that he sustains the pupa sometimes eight or nine months without food, during which it has no apparatus for eating and is shut up in its narrow prison-house. But we will find in our rambles some moths that never eat at all after they cease to be caterpillars."

Just at this point the supper-bell closed this pleasant conversation, but not without the promise on the part of Uncle Samuel that after tea he would show them some more of his drawings illustrating the change from the caterpillar to the pupa state.

CHAPTER III.

"HOW bright and cheerful you all look!" said
Uncle Samuel to his young, earnest and willing pupils as they entered his study after tea.

"Supper is over, you know," answered Mary,
"and we have hurried in after you to hear you tell
more stories about insects, and to see the pictures
you promised to show us."

"I am just looking for them in my portfolio.
They will help you to understand how wisely, and
with what foresight, caterpillars provide for their
safety during that period of their life when they
cannot move about as they once did. Here they
are; I have found them. I made them when in
England to represent the changes undergone by the
common cabbage-butterfly, called *Pieris brassicæ.*"

"The old naturalists called all the butterflies *Pa-*

pilio, because *papilio* is the Latin word for *butterfly*,
and all the moths *Phalæna*, from the Greek word
phalos, a *moth*.   I have an old work on natural history by Stewart, in which this butterfly is called
Papilio brassicæ, the cabbage-butterfly, for *brassicæ*

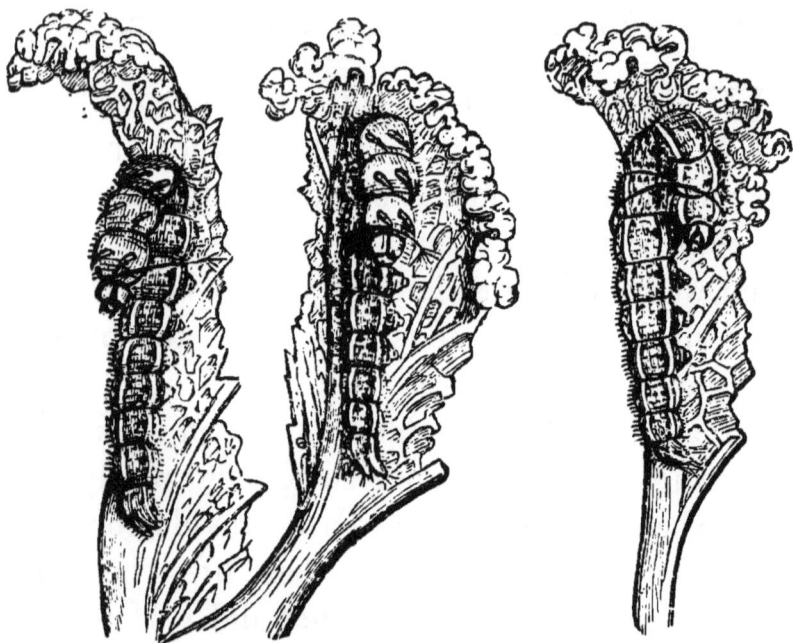

Fig. 9.—CATERPILLARS OF THE CABBAGE-BUTTERFLY
(*Pieris brassicæ*).

means *cabbage*, and the butterfly is so called because
it lives on cabbage.

"In 1801, Schrank, a distinguished naturalist,
made a new classification of the butterflies, and this
butterfly, with other species, was set apart to the
genus *Pieris*.   I have no doubt he derived this
name from an ancient Roman story, which I will
tell you.   As the story goes, there once lived a
king of Mesopotamia called Pierus, who was said

to be the father of the Muses. These Muses were called Pierides, after him, and were the patrons of the poets, goddesses of poetry and music. In the name of this genus, which embraces many very handsome butterflies, the memory of these fabled goddesses is preserved.

"The *Pieris brassicæ* is very destructive to cabbage in its caterpillar state, and feeds also on the turnip. It is abundant in all parts of Europe, but has not yet appeared in the United States. A few years since the *Pieris rapæ*, another cabbage-butterfly of European origin, came into this country, and it is the butterfly that destroyed your cabbage last summer."

"What is the caterpillar doing with a string about its head, and who put it there?" asked Mary, whose attention was attracted more by the pictures than by the historical romance told by her uncle.

"It has great need for that string, and it made it and put it there itself," answered Uncle Samuel. "When the time of the caterpillar's change comes, it selects a suitable place and makes a small hillock of silk, to which it hooks itself by the claws of its hinder feet. It now sets to work to secure a support for its long body while it is passing into its new form. This it does by spinning a silken thread, which it fastens on both sides of its body over the fifth division. The drawings represent the caterpillar in different stages of the act of making this thread. It acts like a creature of intelligence that fully understands what it most needs and is skilled in the art of making silk. It secures its thread in its proper place on one side, then carries its head over its body and secures the other end of the thread against the support on the other side. This operation it repeats until the band is strong enough to bear the weight of the chrysalis. When this work is completed, it withdraws its head from under the band and takes the position shown in this drawing.

Fig. 10.—CATERPILLAR OF THE *Pieris brassicæ.*

" To get into this position it has only to cause its head to slide along the threads near one of the places where they are fixed. In this position it patiently waits for that wonderful change which will convert it into a chrysalis. This requires about thirty hours, during which the caterpillar must be experiencing very strange sensations, if it has any feeling at all. And if it has any thoughts, it must be wondering what is the matter or what is about to become of it.

" When the change is completed, it does not look at all like its old self. Look at this drawing, and you will see the picture it makes when a chrysalis. I have represented it in two positions, in each of which you will see the use it has made of the small tuft of silk and the string it fastened around its body. Strange that when it was a caterpillar it should know so well what it would need after it would cease to be a caterpillar!"

"Poor thing!" said Mary; "how can it ever get out of such a case if it wants to?"

"And what does it do in there?" asked James. "Does it just sleep all winter?"

"Your questions are both very proper, and I will try to answer them. If you could see inside of the chrysalis, James, and observe what is going on there while the insect seems to be at rest, you would think it had not much time to sleep. Why, its wings and legs and head and mouth have all to be made and put in their proper place before it can be a butterfly, and all this must be done during the few months that the insect is shut up in its little pupa-case. So

Fig. 11.—Pupæ of Pieris brassicæ.

you see it must be very busy all this time. And it is busy night and day—so busy that it has no time to eat. Then, just as soon as all the parts of the butterfly are formed and ready for use, it prepares to leave its shell; for insects never lose any time, as boys so often do, putting off until to-morrow what ought to be done to-day.

"Mary wants to know how it can get out of its close prison when it wants to. It understands how, and goes about the work as if it had done it often. It knows that the walls that confine it are not very strong, and as it is loosened from every part of its case it moves itself about at its pleasure. Then, stretching itself a little, the dried and brittle case splits open on the upper part. By repeating its motion and spreading itself a little more, the rent increases until the split extends over the middle of the forehead and back. Then the pieces just over the back open, separating themselves from the other parts to which they were fixed, and the insect, seeing its way to the light made clear, lifts its head and looks out. Little by little it advances, for it must be very careful not to rend or otherwise injure its very tender body and the beautiful, thin, light-colored wings, of which it may justly be proud."

"It seems to me," said Mary, "that the dear little thing would be *so* glad it was going to be free again that it wouldn't stop to think about itself and to take so much pains to get out of its case, but would just fly away at once as soon as it got its head out. I would."

"I suppose it would do so, too," said Uncle Samuel, "if it could; for insects never lose any time in doing their work. But it is not a very easy task to get free from such casing. All its parts—its wings, its legs, its horns, or antennæ— are shut up in special cases, which must be broken open or they must be otherwise withdrawn from them before the butterfly can be free. So you see why this operation requires so much care on the part of the little prisoner.

"At last, by patience and perseverance, the butterfly is free, and poises itself for a while proudly on its cast-off pupal case. What do you think its thoughts must be when it first sees the light, and takes its observations from the outside of its former prison-house, and inhales the perfume of the flowers which are so soon to give it food? If you could only be a butterfly for a little while just when it leaves its coffin, so as to know its thoughts and its feelings, you could tell all about them. But this never can be, so that we can only wonder what the newly-winged butterfly thinks about itself when it first breathes the air of freedom. Here is a picture taken from a drawing made by Reaumur, a French naturalist, which represents a moth just emerged from its pupal case.

Fig. 12.—MOTH JUST EMERGED.

"The wings are folded up inside of the case, so as to occupy but little room; and when the moth first appears the wings are flat and thick, as in the cut. This appearance they have because they are folded and refolded on themselves. By and by the wings grow, and as they

grow they curl up, as is represented by the following cut, taken from another drawing of the same
naturalist.

"Gradually they spread out, assuming by de-
grees their proper proportions, until
they are ready for use. As the wings
expand and harden the whole body
becomes more firm, so that all the
parts are fitted for action at the same
time. In these two drawings (Figs.
14, 15) you see represented the wings
as they appear while they are expanding to full size."

"How wonderful!" said Charlie.
"It seems to me there is a *great deal*
to admire in the different stages in
the life of a butterfly or moth. I never heard of

Fig. 13. — MOTH
WHOSE WINGS
ARE FOLDED UP.

Fig. 14. — MOTH WHOSE
WINGS ARE DEVELOP-
ING.

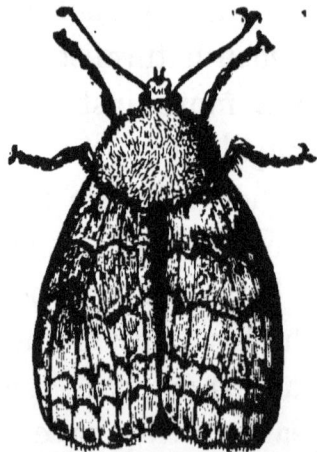

Fig. 15. — MOTH WHOSE
WINGS ARE FULLY DE-
VELOPED.

facts more astonishing. Do they not prove that there must be a great and wise God who made the insects, little as they are, and taught them how to preserve their life and how to do all the strange things they do so skillfully? How careful the caterpillar is to select a safe place in which to stay while it passes into its pupa state! And when it is ready to come forth winged, how wisely and carefully it acts, so as not to do itself any harm while it is so soft and tender, waiting on its rejected case patiently till all its parts harden and become strong and fit for use! I think we children ought to learn a lesson from this fact—not to expose ourselves to bad influences while we are young and tender and our good habits are not strong enough to resist them. If we do, we shall act like the butterfly that will not patiently take its legs and wings out of their special cases, but pulls them out too rapidly, and so breaks and tears them. Is it not very kind in the great God to give to this little insect so much wisdom that, though it never came out of a pupal case before, it knows just how to do it and just when to do it? If he does so much for caterpillars, we ought to believe that he is ready to do much more for us."

"You take a very sober view of things, Charlie," said Henry, "but I think you are right. I believe that if we all had more patience we should succeed much better. Yesterday I tried to separate two leaves of my arithmetic which would always

4

plague me by turning over together, but I did not
go about it calmly and carefully like the butterfly.
I got impatient and tried to pull them apart by
force, and tore them both in two, and the teacher
punished me for carelessness."

"I am glad," said Uncle Samuel, "that you
readily perceive the lessons which these wonderful
facts teach us, and are disposed to govern your con-
duct hereafter by them. If you *think* about the
teachings of natural history, as well as *learn* the
facts it reveals, you will find our study not only interesting, but profitable, and its influence will help
to fit you for acting well your part in life. Always
trust in God as your best and kindest friend. He
takes care of all the creatures he has made, and has
provided them with everything necessary for their
growth and happiness; and if you do not let sin
separate you from him, he will make you infinitely
more happy than any insect that sports itself joy-
ously in the air and sips the nectar from the bloom-
ing flower."

CHAPTER IV.

SPRING had well advanced when one Saturday
morning the children, with their butterfly-nets
in hand, took a ramble over the fields in search of
butterflies. They had some rare sport and capital
exercise while chasing their game and attempting
to catch it on the wing. It was amusing to see
the disappointed look as one and another would
examine his net with the conviction that he had
certainly been successful, and find it empty. The
zigzag course of the butterfly's flight renders it
very difficult to capture. This is its only means of
self-defence when on the wing. The flight of
birds is in a straight line, while the butterfly turns
quickly in its flight, and changes its course so suddenly and frequently that the bird that would
make it its prey is outwitted and the butterfly escapes. So, while their young pursuers were bring-

ing their nets down upon the butterflies in one direction, they quickly changed the course of their flight, and left them to search their nets in vain for their captives.

Mary, seeing one resting on a wild flower, approached it very cautiously, and brought the mouth of her net directly over it. This capture closed the ramble, for they were all anxious to learn the history of their fluttering prisoner. They were not long in reaching their uncle's studio, where he was always ready to talk with them on such subjects.

"I caught it with my little net," said Mary. "It didn't know I was coming, and I just put my net over it, and it couldn't get out. Isn't it a beauty? Oh, poor thing! it wants to get out so badly it will break its wings. Would they bleed if they break, and hurt the dear little thing? What's its name, uncle? I'm so sorry for it. I wouldn't like to be caught in a net and taken away from my home. I wonder what it's thinking?"

"What a wonderful mixture of ideas has filled your little brain this morning!" said Uncle Samuel. "Your sympathy for your beautiful prisoner has almost overcome your anxiety to know its name and history. Your philosophy would make it tell all it is thinking about, and you ask questions too fast to wait for answers.

"By comparing your butterfly with the specimens in my collection you will very easily find out its name for yourselves."

This was very eagerly done, and they all decided that its name was PAPILIO *asterias*. Charlie at once took down his Latin dictionary, and found that *papilio* was the Latin word for *butterfly*, and that *asterias* meant *a gem having the appearance of a star,*

Fig. 16.—THE ASTERIAS BUTTERFLY. (*From Tenney's Natural History.*)

and that it was derived from the Greek word *aster*, a *star.* Combining the two words, he said,

"I understand why it is so called : it is *the stargemmed butterfly.* See the yellow star-spots on its wings."

"Your derivation is very ingenious," said Uncle Samuel, "and you are right about *papilio,* but it is called *asterias* because it was first found feeding on the plants belonging to the order Asteraceæ. The plants were so called because their flowers are *star-like.* The spots on the wing of this butterfly are not *star-like*, as your eager imagination suggests, and so you will have to give up your ingenious theory for facts."

"But what do you know about it, uncle?" said Mary, somewhat impatient with the long discussion about its name, which she could not clearly understand. "Had it to be a caterpillar before it could be a butterfly? If so, what did the lovely little crawling thing live on? for I think such a beauty of a butterfly must come from a lovely caterpillar, if it ever was one."

"Yes, it had to be a caterpillar, like all other butterflies, and it was then, as now, a thing of beauty. If you look on the leaves of the parsley or carrot some time next June, you may find it at home there. It is known best by the common name of the *parsley-worm*. It is very small when it escapes from the egg—all caterpillars are small then—and it is of a black color, with a broad band across the middle and another on the tail, and on the back are little projecting points. It grows very rapidly till it is about one and a half inches in length. One of the strangest facts in this caterpillar-history is that it changes its color at each stage of its growth. When it is full grown, its projecting points and white bands entirely disappear, and the skin becomes smooth and the color a delicate apple-green, with a transverse band of black and yellow spots alternately arranged on each segment."

"How do caterpillars change their color when they grow? And what do you mean by different stages in their growth?" asked Henry. "I should think if they changed their color they would have to change their skin too."

Fig. 17.—CATERPILLAR OF ASTE-RIAS. (*From Tenney's Natural History.*)

"The growth of caterpillars is a very interesting and wonderful part of their history, and this is a very good time to tell you of it. Their skin does not grow with the other parts of their body, like the skin of other animals. It is made to fit the caterpillar exactly; and when it eats so much that its skin is too small and feels uncomfortable, then it has no more use for that skin, and proceeds to cast it off. · This process is called *moulting*, and it is a very delicate and important operation. It requires, also, much time and labor, and is attended with not a little danger. It is, therefore, a critical period in caterpillar-life. The necessity of this change of skin seems to make the caterpillar very thoughtful, and for one or two days before the important moment arrives it ceases to eat, as though there was something very weighty resting upon its mind. But this may be done to reduce the size of the body, and thus render its withdrawal from the skin more easy and safe, or to enable the minute organs, which up to this time carried nourishment to the skin, to withdraw entirely from the skin, so that, being dry and dead, it might be rejected by the caterpillar without danger to its life.

" In the mean time, it seeks some secluded spot in which it can remain undisturbed during the operation. If it lives in society, as the tent-caterpillars do, it retires into its nest and fixes the hooks of its feet during the process firmly in the nest-web. As the time of the change approaches the color declines in brilliancy, the skin withers and hardens, and the juices by which it had been nourished are withdrawn. If you were to look at the insect now, you would see it raising its back into a bow or stretching its body to its utmost extent, as if it were uncomfortable. But I suppose it does not feel pain, and, indeed, the operation may be a very pleasant one. Sometimes you would see it raising its head, moving it from one side to the other for a while and then letting it fall back again. It acts as though it knew that there was a great change going on all over its body which it did not understand, and hence its apparent uneasiness; or perhaps it feels that its old skin is no longer of any use to it, and it has determined at every cost to cast it away.

" Keeping your eyes on the anxious worm, you would see, as the change drew near, the second and third segments increasing in size. Thus the insect stretches the old parts as much and as rapidly as it can. At length the vessels that gave nourishment to the old skin are all withdrawn from it, and a slit appears on the back, beginning at the second or third ring. Now, as the caterpillar feels itself freed from its old and worn-out garment, it presses its body, clad in a new and sometimes a differently-colored skin, through the rent, and gradually withdraws itself from its former covering, leaving the old case, a perfectly-shaped caterpillar sheath, with all its exterior parts—skull, jaws, antennæ, legs, claws and spiracles."

"I found just such a sheath," said James, "in our garden last summer, and I thought it was a dead caterpillar, and I wondered how it had all dried up and yet its skin looked as plump as when it was living. Now I know it was only the cast-off clothes of a caterpillar. How many times does the caterpillar cast off its skin before it is full-grown?"

"Generally three or four times," answered Uncle Samuel. "Each time, after moulting, the caterpillar fasts a whole day to give time for the parts to harden, so that all its organs can do the work expected of them, and then it commences anew to devour the leaves of the plant or tree on which it lives.

"When the caterpillar of the butterfly which you have just caught has moulted for the last time, it comes out of its old skin having a pair of orange-colored horns situated in the first segment just behind the head, which it thrusts out violently whenever it is touched or suspects danger. These V-shaped organs give forth a disagreeable smell, and hence seem to be defensive weapons with which it protects itself from the attacks of a kind of flies which deposit their eggs in the bodies of caterpillars and are their greatest enemies, because their eggs produce little worms which feed on the substance of the caterpillar and kill it. God thus provides a crawling worm with the necessary means of self-defence, and gives it an instinct ever awake to a sense of danger and prompt in the use of organs for its own protection."

"That is wonderful," answered Charlie; "and since God has thought it worth while to provide thus for the safety and self-defence of a caterpillar, we ought to admire more than ever this part of his handiwork. For my part, I am becoming more and more astonished at my own blindness and stupidity in not noticing or studying out the uses of the various organs of insects. They are all so well adapted to their varied and respective uses that they satisfy me that God designed them for those very uses when he made the insects."

"I am glad to hear you say so, Charlie," said Uncle Samuel; "and if you would carefully study the structure and observe the uses of the different parts of insects, you would be confirmed in your present conviction, and would not cease to admire and praise the wisdom of the great Creator, who, notwithstanding the grandeur of his empire, has not neglected or slighted in the least the most insignificant of his works, and who has so wonderfully adapted each of his creatures to the position it occupies in the natural world."

"Tell us, uncle, what it turns into when it quits being a caterpillar," said Mary, who was less concerned about the theology that more mature minds might discover in its structure than in the history of its growth into a butterfly. "Wouldn't I like to see it make a cocoon of silk or dig a grave for itself in the ground! What does it do next, and how does it do it?"

"You could never see it do either of the things you mention, for it never learned how, and never will. It has a way of doing itself up into a chrysalis that it likes much better than either of these ways. Every caterpillar to its taste, you know.

"When its caterpillar-life is about to end, it seems to know that it is not going to die, but that a very great change is about to take place in its life and condition. So you might see it walking about as if in deep thought. It refuses to eat, and withdraws itself from its accustomed places of resort and seeks a sheltered spot—it may be the side of a building, or a fence, or, more likely, the trunk of a tree—where it can prepare itself for its expected change. This is a solemn moment in the experience of the parsley-worm. It has eaten its last breakfast; it will nevermore have any desire to feast on the rich pulp of the parsley or the carrot-leaf. It is about to close its eyes on familiar scenes, that they may enlarge and increase in beauty, and be fitted to look out upon the new glories of higher life.

"Now its instinct acts as it never acted before. New and wonderful powers are displayed. It spins a little tuft of silk as skillfully as if it had been trained in the art, and, as though it fully comprehended the necessities of its new condition, it glues it to the surface on which it rests and entangles its hinder feet in it, so as to fix them securely to the spot. When thus suspended, it makes a U-shaped loop of many silken threads, with its ends fastened to its resting-place. Under the loop it passes its head, gradually working it over its back, so as to support its body, and keep it from falling during the process of transformation. For twenty-four hours after this Nature is very busy changing the form of the caterpillar, and within this time it has actually succeeded in stripping itself of its apple-green skin and shutting itself up in its cosy little chrysalis.

Fig. 18. — CHRYSALIS OF *Papilio asterias.* (*From Tenney's Natural History.*)

"Here is a cut of its new shape and dress. How unlike its former self! It is no longer the active, voracious worm, but a sleeping, motionless pupa, awaiting a still greater change and a life of greater activity. Nor is it without beauty. God honors it still with graceful touches of his pencil. It is pale-green, ochre-yellow or ash-gray in color, and has two short ear-like projections above the head, and a little nose-like prominence on the upper part of the back.

"It remains in this state from nine to fifteen days, and then its skin breaks open and one more of Nature's marvels is seen. A butterfly marked and ornamented just like your beautiful captive issues from it, clings to the vacated shell till its crumpled and drooping wings have stretched themselves to their fullest extent and become dry and ready for use, and then, clad in its new robes of beauty, and gifted with new tastes and capacities for higher and purer enjoyment, it bids adieu to the faithful shell which had so lately protected it, and with graceful ease, as though it had always been winged, it flies away to find new companions, and to participate in new pleasures and sip the nectar from the richly stored flower-cup."

"What a beautiful butterfly it is!" said Henry. "Let me see if I can describe it. Its chief color is black. It has a double row of eight yellow spots on the back of its abdomen, and two rows of similar spots on each side, and two yellow spots just behind the eyes. The wings have two bands of yellow spots between the two yellow bands, and near the hinder angle an eye-like spot of an orange-color with a black centre. The most of the spots on the under side are tawny orange. The wings expand about three and a half inches."

"Your description is very good and shows accuracy in observation—a very important trait in a young naturalist. The cultivation of this power of the mind is one of the results of the study of entomology, and this fact should recommend it as one of the best means of thorough *mental training*. Note closely every mark in an insect, observe its size and position, the form and outline of each wing, the number of the legs, the length and form of the antennæ, the proboscis, if it has any; these things fix the class and the species of the insect, and will enable you always to recognize individuals in the insect communities as you recognize old acquaintances.

"During the month of July you can see these butterflies in great abundance; but sometimes, as it was with the specimen just captured, they appear earlier in the season. In July and August they lay their eggs on various plants, placing them singly on different parts of the leaves and stems. This is done, perhaps, for the greater security of the young, and, it may be, that the interests of the young may not be brought into conflict when they are foraging for their food. The young caterpillars feed on flowers and seeds, as well as on leaves, so that they are often very destructive of many cultivated plants and a great enemy of the gardener. Harris recommends the gathering of them by hand and crushing them as the most effectual method of banishing them from the garden."

"I think that would be an effective mode—at least, for the crushed worms," said James. "But isn't it wrong to kill a worm that is so beautiful, and that grows up into such a beautiful butterfly?"

"Its beauty won't save it," quickly answered Charlie, "if it is mischievous and bad. There is no goodness in mere beauty. Sometimes the most beautiful persons are the most vicious. Wicked men often lose their lives because of their wickedness, and why shouldn't worms? If a worm destroys what man has planted for his own use, I think he is justified in defending his own property by killing the worm. Then, if man is 'lord of creation,' he has a right to assert his authority over caterpillars, and to take their life if they destroy the vegetables on which he lives."

Our young naturalists, in their attempt to moralize, would doubtless have become terribly involved in some knotty questions of religious duty toward the lower animals, had not their conversation been suddenly cut off by the arrival of their little city cousin, who had come to spend the season with them. The beautiful butterfly and its right to life and the pursuit of happiness in its own way, as well as the higher right of man to protect his food, were promptly dismissed from their thoughts that a hearty and joyous welcome might be given to their long-expected friend.

CHAPTER V.

"HAVEN'T we had a good time, though!" said
Mary as a kind of apology for their absence from their uncle's study. The whole household had been made very happy by the arrival of
little Bertha, who was not much older than Mary,
and who had never before been in the country. To
her everything was new—the garden, the fields, the
woods. She had often seen Uncle Samuel's collection of butterflies and other insects, and had learned
from him the names of some of their organs; but
she had never seen so many beautiful insects flying
in freedom over broad meadows and by the side of
running brooks.

When, therefore, she had gratified her curiosity
about the things that were new to her in the house-

and the yard, she was very willing to roam with her cousins over the fields and in the woods to gather moss and wild-flowers, and to give chase to the sportive butterfly.

This evening our little group of naturalists, recollecting that their last talk with their uncle was very abruptly broken off, and supposing that he had some more interesting stories to tell them about butterflies, found their way into his study. They had not forgotten what he had told them ; and while resting on the grass by the side of the stream which flowed so gently through the meadow, they had told it all over to Bertha, exciting in her a great desire to hear more about insect-life, and preparing her for listening intelligently to everything her uncle might have to tell them.

" Do tell us some good stories about butterflies," said Bertha, whose zeal for such stories had been wonderfully wrought up by the enthusiasm of her cousins.

" With great pleasure, madam," answered her uncle. " I am always glad to show the little ladies who live in the city what wonderful things are to be found in the country, so that they will love the country more, and love and worship the great God who made the country so rich and beautiful, and who fills the air and land and water with such wonderful things of life as are to be found in the insect world. For your better entertainment I will call in the aid of my microscope, and

5

show you wonders which otherwise no one could ever see."

Taking the butterfly in his hand and calling his attentive listeners near him, he directed their attention to the coloring of its wings.

"Notice," said he, "the wings of this butterfly. See how accurate its markings are, how delicate the shading, one color seeming to glide into another. Could any painter excel or equal the touches of Nature's pencil?"

"Are the wings really painted, uncle?" asked Mary. "Oh, there! I have rubbed out one of its spots. It was a yellow one, and now the wing looks as clear almost as glass just where I pressed my finger upon it. I do not think that Nature's paint sticks as well as man's. If it did, it wouldn't rub off as easily as it does."

"Look at your finger, Mary," said James; "it has dust on it. I think that dust came off the butterfly's wing. It looks yellow, too. It is the *paint* taken from the wing; that's what it is."

By this time their interest had greatly increased, and our young friends were very anxious to know what kind of coloring-matter Nature made use of in painting the butterfly's wing. In the mean time, Uncle Samuel had adjusted his microscope, and was ready to satisfy their curiosity by showing them some things more wonderful than any with which they had yet met.

"I am not astonished," said he, "at your anxiety

to know where Nature gets her paint and how she puts it on the wing of the butterfly. I ought rather to say the *God* of Nature, for I want you always to remember that there is no such thing as Nature independent of the all-wise Author of Nature— God himself. When, therefore, I speak of Nature doing anything, I mean God working in and by means of the laws of Nature. Now, if Mary will only press her finger that is marked with the dust she unintentionally took from the wing of the butterfly upon this small piece of clear glass, I will soon show you what that dust looks like and how it is made to give color to the wing."

Mary did so; and when the slip of glass was held up to the light, nothing was seen but specks of fine dust. Great was the interest manifested by our young friends in their prospective discovery of what these specks were. As soon as the glass slide was put in its place, so that the scales could be seen with the microscope, one after another, commencing with the youngest, was permitted to look at them. It would have done you good to have seen the expression of wonder and delight which lighted up the countenance of each as they respectively gazed upon the dust enlarged into variously outlined scales. Bertha's astonishment was so great that she suspected it to be one of her uncle's tricks.

"I don't believe a word of it," said she; "you are only making believe. I know that is dust. You just put something else in there, and want to make us believe it came from the butterfly's wing. If there were anything on its wing like that, why we could see it. Then they're so large, and look just as if they were cut out with the scissors by a pattern. I just *know* you're playing some trick on us."

"Well, to convince you that the microscope is telling you the truth, and that I am not playing any trick on you, look closely at this glass slip," said Uncle Samuel as he held the slide in his hand, having removed it from the microscope. "What do you see?"

"Why, nothing but dust from Mary's finger," answered Bertha.

"Now look into the microscope and tell me what you see."

"I do not see anything at all but the lower part of the microscope, and that very dimly," said she.

"Now carefully observe that I am putting under the object-glass that very slide that you say is covered with dust. Put your eye now to the eye-glass."

"Oh, now I believe it all; and how beautiful, how wonderful! Who would have thought it! Some of the scales are yellow, and some are dark-colored. They are notched at one end, some have three teeth like a saw, some two, and some four or more. There is one long and narrow; some are shaped like a heart at the end that is not notched, and out of the groove in the heart is a stem like an apple-stem, and some have the stem without the heart."

Fig. 19.—MAGNIFIED SCALES OF THE *Papilio asterias,* AS SEEN BY BERTHA AND HER COUSINS.

"I think it is wonderful," said Henry as he took his place at the microscope. "Does the color come from these scales? It looks as if it did, for that part of the wing from which they were taken is almost as clear as glass. I wonder how they are fixed on the wing? It must require great skill to fasten such dust-like particles on so thin and delicate a thing as a butterfly's wing."

"These scales hold all the coloring-matter of the wing," said Uncle Samuel, "and there is wonderful diversity of shape among them. You see that no two of them are alike, just as it is with the leaves of an oak tree, while the combined effect of the natural mosaic on the butterfly's wing is as striking and beautiful as the outline of the oak is majestic and grand."

"Why are the scales not alike, and why do they

not all have the same number of notches?" asked
James.

"Because each scale is designed to fit a particular
place, and it must have the particular outline and
the special tint which adapt it to the portion of the
wing it occupies. Any change in the form of the
scales, by the lengthening or shortening or removal
of a single tooth, would mar the beauty of the
marking and the perfection of the shading of the
spots. In a single scale there is nothing which displays any great skill, and there would be no great
difficulty in drawing a more accurate mathematical
figure and one that we should think more beautiful; but when we look upon the rich coloring of
the wing, and the regularity of the spots by which
each species is distinguished, and the perfect blend-
ing and shading of the various hues, we cannot help
admiring the exquisite taste and admirable skill
displayed in the construction of this wonderful
mosaic."

"Mosaic! What do you mean by that word,
uncle?" asked Mary. "Is it anything about Moses?"

"Mosaic is the name of a kind of inlaid work,
like patchwork, made of very small pieces of hard
substances, such as glass, marble, stones and gems,
carefully laid together and cemented, so that the
whole work will represent some beautiful picture.
These small pieces are of various colors, are generally cubical in form, and when put together look

like stitches of different colors in worsted work. It does not receive its name from Moses, but from a Greek word which means *polished*, because the pieces are all polished after being fitted to their places perfectly. Now you can see, by means of the microscope, what kind of pieces are inlaid in the butterfly's wing to make up the beautiful picture which is so much admired. These pieces are very small and very carefully fixed in their places. They are so small and so closely put together that the number of them on a single wing is inconceivably great. In a mosaic, human skill has put as many as eight hundred and seventy separate pieces in a square inch. This is regarded as a great triumph of art; and so it is. What, then, shall we say of the art that puts, in the same extent of surface, the astonishing number of two hundred thousand separate pieces, all in their proper places, without a single mistake to mar the beauty or destroy the symmetry of the picture? If we admire the ancient mosaic and wonder at the skill and patience of the man that constructed it, we ought to adore the wisdom and skill of the great Jehovah, who works with matchless art even on a butterfly's wing.

"The most perfect workmanship of the most celebrated artist is but clumsy bungling when compared with the skill of the Divine Architect. How amazing are His works, and how much of their beauty and grandeur is brought to our knowledge by means of the microscope! The world is full of pictures of beauty—in the gorgeous tints of the delicate flower, in the brilliant plumage of birds, and in the exquisite mosaic which displays its glory on the wing of the butterfly and moth."

"Uncle," said Mary, "Bertha says she thinks the scales must be put on the wings with very poor paste, they rub off so easily. Just look here! Henry has rubbed nearly all the scales off one of the wings. Do tell us how they are fastened to their places."

"That I will do with great pleasure. But you must believe what the microscope tells you, for I cannot make my answer plain to you without its aid. I have taken the piece of the wing nearly stripped of its scales by Henry and placed it under the microscope, so that you can all see for yourselves how the union of scale and wing is effected.

"While you are examining the wing I will explain what you see. The part of the wing that is scaleless is clear and marked with ridges which appear like small dots or holes. Into these ridges the stems of the scales are inserted, and thus are fastened to the wing. The ridges are in rows, and so close that the scales overlap like the shingles on the roof of a house, so that you might

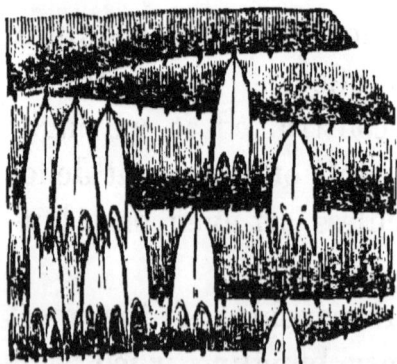

Fig. 20.—Scales on Butterfly's Wing.

say that the wing is roofed with scales. The wing is very thin, and yet on each side of it you will find these ridges, each one holding its own scale in its proper place. No scale ever misses the ridge to which it belongs and exchanges places with some other one. If such a thing should occur, the picture would be spoiled, and the butterfly would be so far deformed."

"I can see now," said Charley, "why the scales differ so much in size and outline. Like the different pieces of a mosaic picture, each scale is made for the place it has to fill, and is cut and painted just as it is required to be. The stem must be made to fit the socket that holds it to its place. If it were too long or too short, it would not answer the purpose. The other day I saw men at work on a bridge which required a great many pieces of iron, which were intended to be made to fit exactly the places in which they were designed to be put. It happened, however, by mistake that several of them were *just a little* too long, so the workmen had to cut a small piece off each one with chisels and files before it could be used. I think that would be the way with the stems of the wing-scales if any of them were too long. They would be of no use till they were made to fit. And the notches would have to be cut just to suit the picture; and so they are. I understand it all now, and it is very wonderful. I have often looked at butterflies with admiration, but I never before knew that there was so much mechanical skill displayed in the formation of their wings.

"I read in ancient history," continued Charlie, "about a man called Epicurus, who lived about three hundred and forty-two years before Christ, and who taught that everything in the world came by chance. I don't think that, with all his knowledge, he ever knew how the scales were fastened on a butterfly's wing, or he would have been compelled to admit that blind chance could never do such work. It seems to me that the whole structure, painting and use of the wing must have been in the mind of God before ever a butterfly was made, and that it was made according to this prearranged plan."

"Some of our wisest men have seen this subject in the same light," said Uncle Samuel. "It is well to look 'through nature up to nature's God.' We should judge of the wonderful mechanism seen in the structure and uses of the different parts of animals as we do of the products of human genius and skill. If we were to see a steam-engine for the first time, and observe how well it is adapted to the work it is required to do, we should reason that the inventor *meant* it to do that very work, and we would not call in question the fact that it was built according to a preconceived plan. So the very existence of the engine would convince us that somebody existed before the engine did, to whom it was indebted for its structure and the power of acting with such apparent wisdom. Ought we not, also, to take for granted, from the exquisite mechanism of a butterfly's wing, the pre-existence of a Being capable of conceiving the plan of its structure, and of putting the parts together for the very purpose for which the wing is to be used? There is such a being, and that being is God, the self-existent Author of all things."

Bertha, to whom this long argument seemed foreign to the subject, because it was not comprehended clearly by her, arrested further discussion concerning the origin of things by the question,

"Are the scales of all butterflies just like those we have seen?"

"In general form they are; as the leaves of different kinds of trees are so similar that you can tell a leaf when you see it, though it may differ from all the leaves you have ever seen. And now, since you have called my attention to the subject, I want to show you how God has displayed his taste in varying the outline and special forms of these very minute parts of butterflies and moths. In all the forest you cannot find any two leaves precisely alike, and yet there is such a family likeness in the leaves of any one species of tree, as the oak, that you can tell an oak leaf from the leaf of the pine, walnut, hickory, beech, or any other kind of tree in the woods. So I am inclined to believe that a careful study of the structure of the scales of the different species of butterflies would enable any one to tell the difference between them, and when he saw the

scale to name the kind of butterfly to which it belonged.   It is because these scales are so small that
they are not so generally and accurately known as
the leaves of trees, or the wool of different varieties
of sheep, or the hair of different kinds of animals."

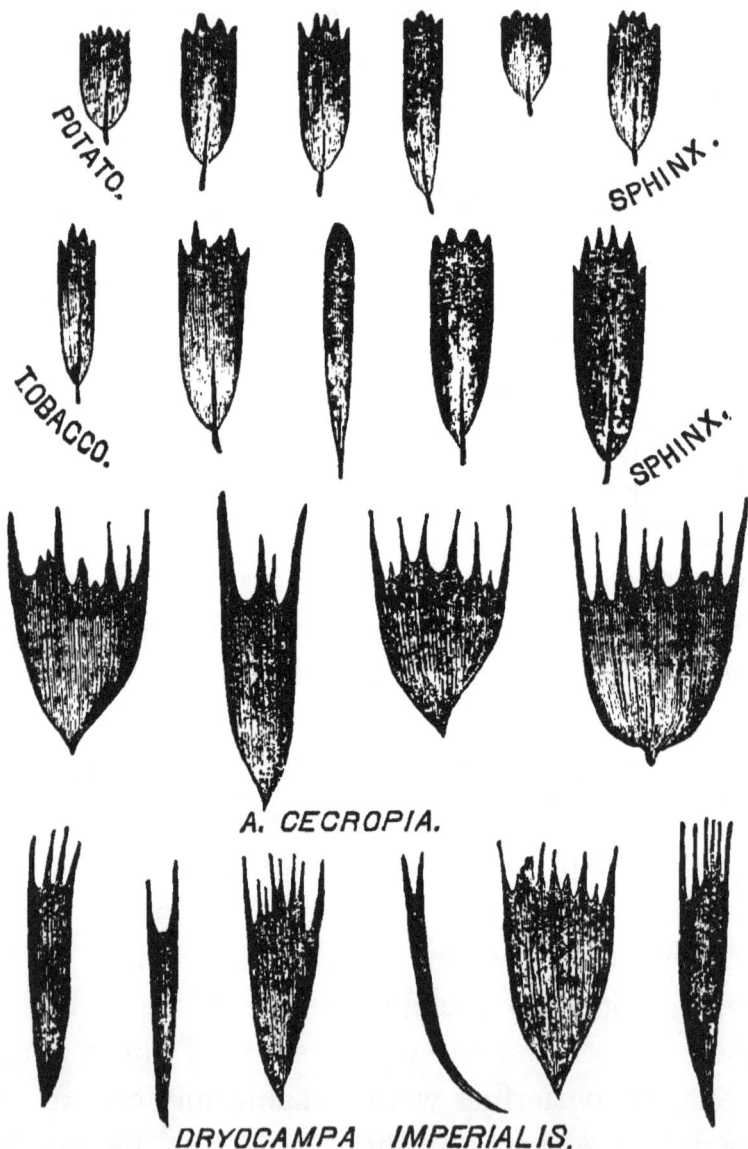

Fig. 21.—MAGNIFIED SCALES FROM THE WINGS OF MOTHS.

SCALES FROM WINGS OF MOTHS

SMERINTHUS GEMINATUS.

FIG. 22

SCALES FROM WINGS OF BUTTERFLIES.

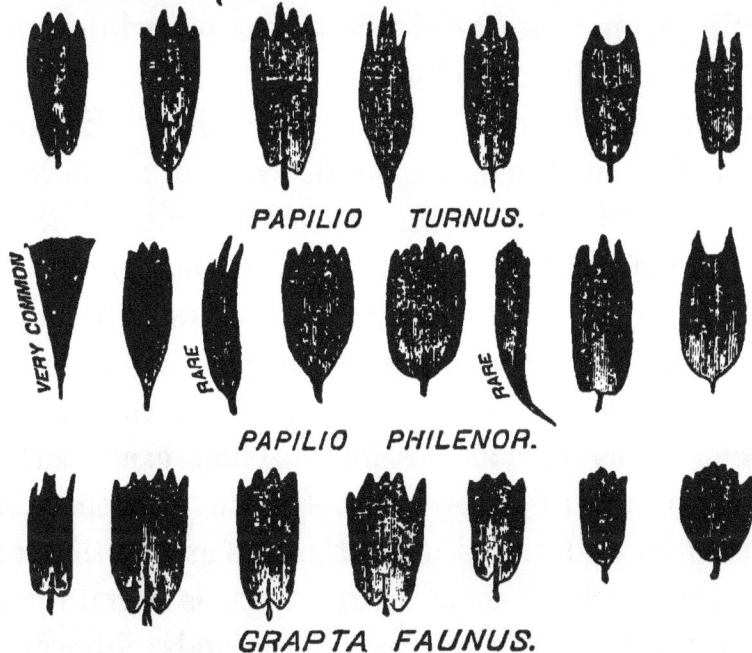

PAPILIO TURNUS.

VERY COMMON

RARE

RARE

PAPILIO PHILENOR.

GRAPTA FAUNUS.

Fig. 22.—MAGNIFIED SCALES FROM THE WINGS OF BUTTERFLIES.

"While studying this subject I made drawings from the microscopical views of the scales of a number of butterflies and moths, which, if you examine closely, you will discover differ from each other just as the leaf of one species of the oak differs from that of another species.

"Look carefully at these drawings, and see if you can discover any characteristic differences between the scales of butterflies and those of the moths."

Charlie, who had a keen eye and was generally accurate in making observations, replied,

"I notice one prominent distinction. Many of the scales of the butterflies end with a stem projecting from a heart-shaped depression, while none of the moth scales end so."

"You will notice, also," said Uncle Samuel, "that the notches of the moth scales are generally sharper and more numerous, and some of them much longer, than those of the butterflies. I should have told you that these scales were taken from the same part of the corresponding wing of each specimen examined, so that the comparison might be as fair as possible. Their examination gives ground to believe that specific differences in butterflies and moths affect the size and outline of the scales which cover their wings, as certainly as the same differences in trees and animals with which we are familiar affect the leaves of the one and the hairy covering of the other. Observation and analogy sustain this conclusion, and confirm our faith in the permanency of species, and prove the idea that a whole species of butterflies can change their color in imitation of the color of another species to be absurd and contrary to fact. But it is time to close this long talk. Come to my study to-morrow evening, and I will tell you more about the structure of the butterfly."

CHAPTER VI.

"HOW glad I am," said Uncle Samuel, "to see
your smiling faces this evening! They tell
me that all has been well with you at school and at
home. A cheerful heart and a good conscience
greatly help the student to acquire knowledge, and
I have no doubt I shall have attentive listeners to
what I am about to tell you."

"Indeed you will," said Mary. "We have been
ever so good to-day. Bertha likes our school so
much; there is such a large yard to play in, and we
have seen so many pretty butterflies. We told our
teacher some of the stories you told us, and he liked
them ever so much."

"Yes," said Bertha, "and he said that he could
hardly believe all we told him about the wing of a
butterfly being covered over with painted scales,

almost like a fish. But we told him we *knew* it was true, for we saw them through your microscope. It seems to me that a teacher should know everything, so that he could tell the scholars nice stories about what God has made in the world. But go on, uncle, and tell us more about the butterfly."

"Before I proceed, I think you ought to learn and understand the name of the order of insects to which butterflies belong. The names of the orders of insects are chiefly taken from the peculiar structure of their wings. Now, if you were going to give a name to the order to which butterflies belong, what would you call them?"

"I would call them 'scale-winged' insects," answered Charlie.

"So they are called by naturalists. But as scientific names are taken either from the Latin or the Greek, they have gone to the Greek and taken the word that means *scale** and another meaning *wing†*; these they put together and make the word LEPIDOPTERA, and this is the name of the order to which butterflies and moths belong. This order is divided into two *sections* or *classes*—BUTTERFLIES and MOTHS. The first of these embraces the day-flying Lepidoptera, and the second those that fly chiefly in the evening and night. These classes are similar in some respects, but differ very much in the time of their activity and their rest. We shall find some moths before we are through with our

* Lepis. † Pteron.

rambles, and then you will learn some very interesting facts in their history.

"Our special study this evening will be the eyes of butterflies; and to enable you to understand how much God has done for them in this respect, I have placed before the object-glass of the microscope a

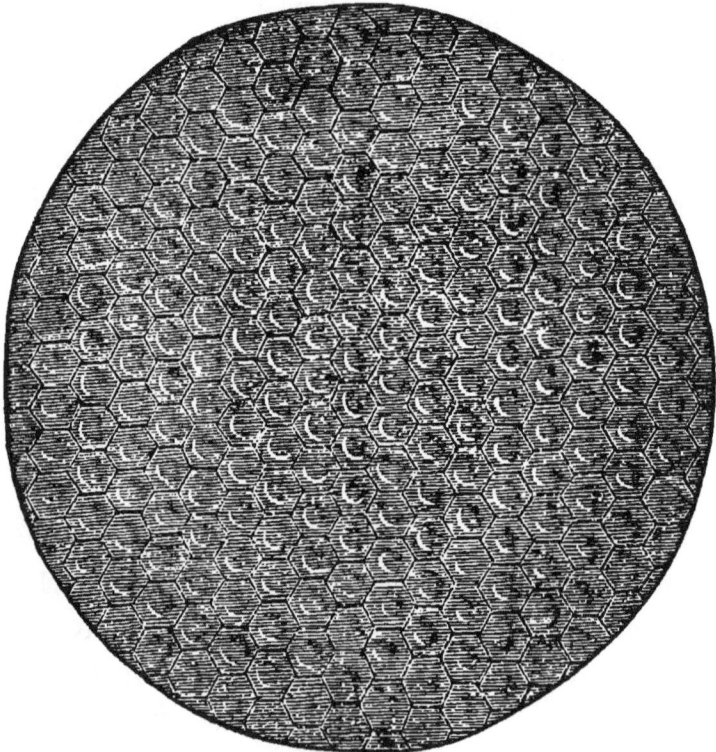

Fig. 23.

slide containing the eye of a fly, and I use a lens magnifying one hundred and fifty diameters, which makes it appear very large. I use this slide because it is convenient, and the eye of a butterfly closely resembles that of a fly. Come and see how wonderfully it is constructed."

"How beautiful!" said Bertha. "It is just like the most beautiful lace; each division has one, two, three, four, five, six sides."

"I think it looks just like a piece of honey-comb," said Henry, "for the cell of a honeycomb has six sides. If every one of these little cells is an eye, I should think the butterfly would see a great many views of the same object."

"You have *two* eyes," said Uncle Samuel, "and yet you see but *one* view of the same object; why may not a butterfly have more eyes than you, and see objects single just as you do? The rays of light that pass from an object through the lenses of your two eyes come together and make but one image on the nerve of vision; so each one of the many lenses of the eyes of the butterfly receives rays of light from the same object, which meet and form but one image on the nerve of vision. The butterfly sees objects in nature, not multiplied, but just as you see them, and as they really are."

"Does each one of these six-sided divisions represent a single eye?" asked James. "If so, how many eyes has a butterfly?"

"Each eye is compound, and is made up of eyelets (small eyes), each containing two lenses through which the light passes, and is as perfect an instrument of vision as your eye is. The six-sided division is called a *facet* or small face, and represents one of these eyelets. A butterfly has as many distinct small eyes or lenses as it has facets. Now, these have been counted, and found to number seventeen thousand three hundred and twenty-five for each eye, making the whole number of eyelets thirty-four thousand six hundred and fifty, united in two large compound eyes."

"Now, uncle!" exclaimed Mary.

"You are just making that up to surprise us," said Bertha. "How can such a little thing have so many eyes?"

"I do not wonder at your astonishment," answered Uncle Samuel, "and that you are almost ready to doubt what I say. But you know that all that God does is astonishing to us, while everything is simple and plain to him. He can as easily give to one of his little creatures thirty-four thousand six hundred and fifty eyes as give you two eyes. You notice that each eye looks very large, and stands out like a little globe on each side of its head. The round surface is hard, so that it can resist injury, and is made up of the divisions you see through the microscope. Each of these divisions admits the rays of light and conducts them to the nerve of sight, called the *optic nerve,* so that there is no confusion in the insect's vision. It sees objects as distinctly as we do, and it may be that small objects are seen even more distinctly.

"Here is a picture that will enable you to understand more clearly the interior structure of the eye. *A* represents a section of a butterfly's eye, *B* a portion more highly magnified, showing the facets

and their transparent pyramids surrounded with a
dark coating. These pyramids meet at *A*, where
they are joined to the
optic nerve."

"How *very* won-
derful!" said Charlie.
"It would seem as if
God had displayed
more mechanical
skill and wisdom in
the structure of the

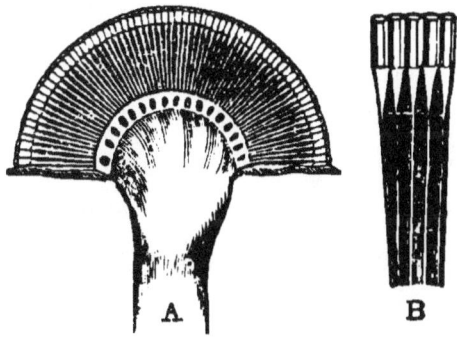

Fig. 24.—SECTION OF EYE OF A
BUTTERFLY.

eye of a butterfly than in that of man. The manner
of his painting the wing and the taste shown in its
markings are wonderful, but the eye surpasses the
wing. It seems so strange that it should have thirty-
four thousand six hundred and fifty eyes. I do not
see why a little insect like a butterfly could not have
done just as well with two eyes as I can. It makes
me think of the words you read in the Bible this
morning: 'O Lord, how manifold are thy works!
in wisdom hast thou made them all; the earth is
full of thy riches.' But it seems to me he has lav-
ished his goodness unnecessarily upon the butterfly
in giving it such a wonderful eye."

"Our eye is under our control," answered Uncle
Samuel, "so that we can move it in any direction
we please, and both eyes move in the same direction
at the same time. We can look up or down or on
either side. Besides, we have power to move our
head so that we can look behind us and directly

above us with ease. The butterfly cannot turn its
eye in any direction, nor can it move its head in all
directions as we do, without moving its body; so
these defects are wisely removed by the peculiar
structure of its eye. The position of the eye in
the head, and its numerous facets, or little faces,
each serving as a separate eye, enable it to see
objects on all sides at once. The chameleon has
power to turn its eyes in opposite directions at the
same time, and thus to look at objects on both sides
at once. Why may not the butterfly with its wonderful eyes not only see objects on both sides, but
also before it, at the same time?

"You know how sharp-sighted the butterfly is.
When you steal up behind it as noiselessly as a cat,
and try to catch it with your fingers, it sees your
hand before you can reach it, and makes its escape.
This power of keen and all-sided vision is given to
this innocent and harmless little creature as a
means of self-preservation. It has no horns to
fight with, no sting to wound its enemy, and no
mouth to bite with. All it can do for its own protection is to use its wings promptly and in time to
get out of the way of danger. Do you not see,
therefore, not only the wisdom, but the great kindness, of God in providing for it such marvelous
eyes, and thus enabling it to see its enemy approaching, and to make its escape in good time?
Is there any unnecessary lavishing of His power
and goodness on this insect?"

Charlie listened with intense interest to his uncle's long defence of the wisdom of God as displayed in the structure of the butterfly's eye, and acknowledged its force, admitting that God knew best what were the wants of his creatures and how to supply them.

James was specially pleased with the idea of seeing on all sides at once, and said,

"Isn't it nice, though? It seems to me that the butterfly is better off than we are, even if it cannot move its head about. Sometimes men get run over by wild horses because they cannot see what is coming up behind them. I believe I'd rather have an eye like a butterfly's."

"I wouldn't," said Henry, "because it has no eyelashes to keep out the dust, and to shut out the light when it is too strong, and I think it would get all covered with dust sometimes, and the sun would often be so bright as almost to make it blind. Don't you think so, uncle?"

"If its eye were soft and tender, as our eye is, it would need some covering to protect it from the dust and rain and bright sunlight; but it is not. It is very hard—almost as hard as glass; and if any dust falls on it, it does not remain, nor does it injure the eye; and it is so constructed that the sunlight cannot be too bright for it. You know that the eagle can look right in the face of the sun and receive no injury, and why should not the butterfly's eye, like the eagle's, be so suited to the light that the brightest glare of the sun's rays could not make it blind or produce pain?"

"You said that this slide contained a part of the eye of a fly," said Bertha. "Has a fly as many eyes as a butterfly?"

"The house-fly has only four thousand eyes on each side of its head, but it is just as happy with only eight thousand eyes as the butterfly is with thirty-four thousand six hundred and fifty. Every insect is provided with just as many eyes as it needs and can use well. The dragon-fly has twenty-five thousand and eighty-eight. Some beetles have fifty thousand. The silk-worm has twelve thousand four hundred and seventy-two, and the ant, little as it is, has one hundred eyes. The eyes of different species are adapted to their manner of seeking their livelihood. Those whose food is near them always have small and flattened eyes, while those that seek their food at a distance have large eyes that are very round; and for a similar reason it is supposed that the eyes of the males are larger than the eyes of the females, because they have to search for their companions.

"Besides these wonderful eyes, which are called compound eyes, the butterfly has three simple eyes, called *stemmata* (or *ocelli*), placed above the head a short distance behind the antennæ. With these eyes they can only see objects that are but a short distance from them."

"I don't see what need the butterfly has for any

more eyes, when it has so many on each side of its head," said James.

"It is observed," said Uncle Samuel, "that most insects that have these simple eyes feed on the sweets of flowers, and some suppose that they enable them to distinguish the different parts of the flower on which they rest. Their large and beautiful compound eyes may be given them for the special purpose of seeing the approach of their enemies and enabling them to find the flowers on which they feed; while the simple eyes, having the power of closer vision, are needed to enable them to discover the food which the flowers contain."

"I think I see it all," said Charlie. "The butterfly needs not only eyes that see at a distance, but it must also see close objects. So a few single eyes are given it for use when it takes its meals. Just as it is with you, uncle. When you wish to read, you put on a pair of glasses with which you can see near objects clearly; but when you look at objects at a distance, you put on another pair. What the wisdom of man has done for you in inventing glasses of different powers of vision, God has done for the butterfly. God did not do this for man, because He gave him an inventive genius by which he could make provision for himself whenever he needed it. He makes up for the lack of genius on the part of the butterfly by providing for its wants Himself."

"What are these long horns for, uncle?" asked

Mary, who was more interested in examining the different parts of the insect than in the moral reflections of her elder brother.

"I will tell you something about the horns of the butterfly after my return from the city. It is late now, and your lessons require a little more study before retiring for the night. You see how useful its eyes are to this beautiful creature, and how faithfully it uses them. Always open, it sees its enemies when at a distance, and makes good its escape. Like the butterfly, keep your eyes always open. You are surrounded by enemies; be watchful, and escape in time the danger threatening you. This is your best means of self-protection. If you close your eyes, so that you cannot see the approach of temptation, it will find you off your guard and lead you away before you are aware of your danger. As it looks after its food, so use your eyes in looking after knowledge. Go where it is to be found as the butterfly goes to the flower. Drink it in, just as the butterfly drinks in the sweet juices it finds in the flower-cup. If you use your two eyes as faithfully as the butterfly uses the many eyes God has given it, you will be much better off than it is, or ever will be. It is because little boys and girls keep their eyes shut that they grow up to be unlovely and ignorant."

CHAPTER VII.

THE lesson taught at their last interview by
Uncle Samuel was not forgotten by his nephews
and nieces during the next day when they took their
places in the school-room. They were very diligent in study, and consequently answered promptly
and intelligently the questions asked them by their
teacher, so that at the close of the day they were
commended for their good deportment and their
perfect lessons. They had kept their eyes open and
tried to make the best possible use of them; and
when they returned home, their great good humor,

91

which always springs from a sense of doing one's
duty well, brought sunshine and joy to the hearts
of all in the household. So it was the next day
and the day following, during which Uncle Samuel
was absent.

Their anxiety to hear more about the butterfly
was so great that they awaited his return with some
impatience. So, when at the close of school on the
third day of his absence they learned that he had
returned, they could scarcely wait till after tea for
another talk about the butterfly, such a craving appetite had they for the more wonderful food which
their uncle supplied them.

"Uncle," said Bertha, "you said that the butterfly had no horns to fight with, but here are two
horns; and though they are not like the horns of a
cow or an ox, they look like little clubs which the
insect might use to beat away its foes. Don't they
fight with them?"

"Your question is a very natural one for a little
thoughtful girl as you are, and I will try to answer
it as well as I can. These horns, or *antennæ* as
they are called, are very stiff now because this butterfly has been dead several days; but when it was
alive, they were very limber, and would bend in
every direction. The insect had power to move
them just as it wished, but it could not strike another insect with them as a boy would strike with
a club. If it could, the stroke would not hurt its
enemy nor drive it away, but would injure the in-

sect itself. They cannot, therefore, be used as a means of self-defence.

"But before I tell you how the insect uses these horns, I want to tell you something more about their name and structure. As to their name, you call them *horns*, and this is their common name; but as they are not used like the horns of animals, naturalists have named them *antennæ*. Now, if Charlie will only tell us what this word means, we will try to find out why such little horns have been honored with such a learned name."

"*Antennæ* means 'sail-yards,'" said Charlie; "we had that very word in our Latin lesson to-day. But I do not see why a butterfly should have a sail-yard on each side of its head. It seems to me that naturalists have very strong imaginations if they can see any connection between sail-yards and these long horns. Do they help to guide the insect when it sails through the air?"

"It is believed that this is one of the uses to which they are put, and perhaps the one which has originated their name. The June beetle, when it loses its antennæ, cannot direct its flight through the air, but flies about in a senseless manner. The Cecropia, one of our largest moths, notwithstanding it has very large wings, cannot guide its motions when its feather-like antennæ, which rise so proudly above its head, are gone. Or they may be so named because they are usually long and slender like the yards of a ship; and in some cases, as the antennæ of many moths, they have branching feathers on opposite sides, so as in some measure to resemble sail-yards with the sails unfurled. They are sometimes called *feelers*, because of their supposed use as organs of touch, but the true feelers are different organs called *palpi*.

" The antennæ of butterflies are long and made of very small pieces jointed together, and end in a knob, so that it looks very much like a club. (Fig. 25.) To express this resemblance they are sometimes called *clubiform*—that is, like a club. The club that Hercules—a very strong, large man, and a great fighter, that lived a great while ago—used was called in Latin *clava;* so some people call them *clavate*. Thomas Say, who has described a great many butterflies, calls the club at the end of the antennæ *conic-ovate* —that is, like an egg-shaped cone. I don't think this is a very appropriate descriptive word, though it *is* high-sounding.

" The antennæ of moths are unlike those of butterflies, and are of various shapes. (Fig. 25.) Sometimes they look like little feathers set in the head. Sometimes they spread out gradually in the middle and taper toward the upper end, terminating in a hook. Their different shapes are too numerous for me to describe in full. You will observe them as you become familiar with different species.

" One of the most interesting facts in the structure of the antennæ of insects is the number of the joints of which they are composed, to which I

might add the manner in which they are put together. Those of the honey-bee, short as they are,
have twelve joints. But here is the microscope. I

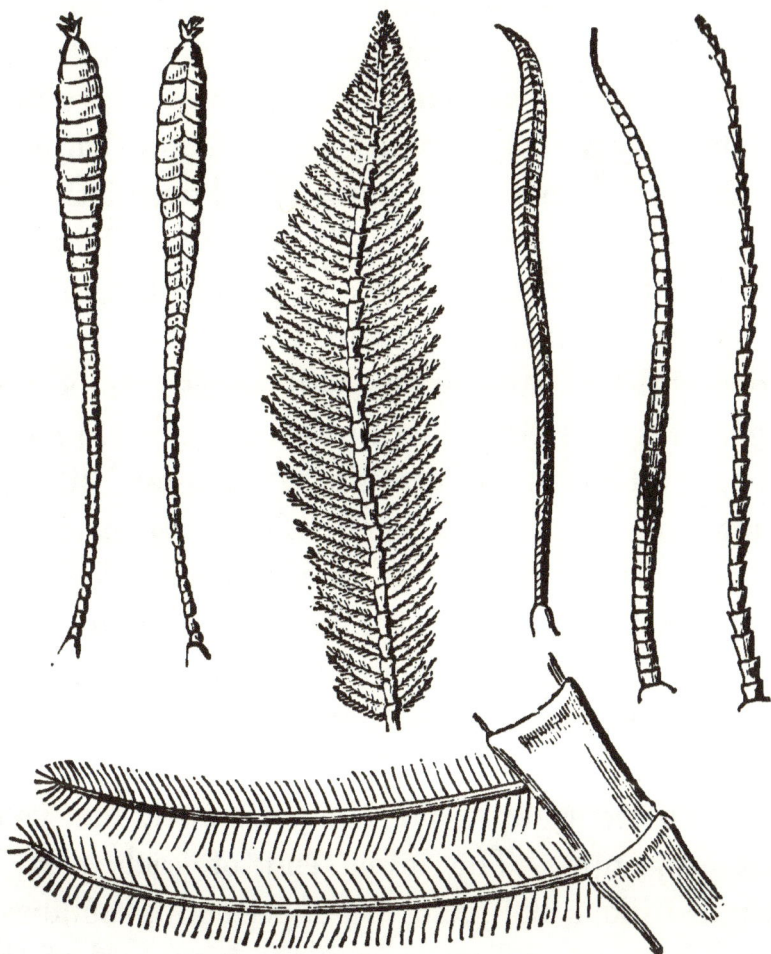

Fig. 25.—ANTENNÆ OF BUTTERFLIES AND MOTHS.

will place before the object-glass an antenna of a
pieris butterfly, and you can examine and count for
yourselves."

This antenna was examined with great interest by
each one of our young naturalists, and the number

of the joints counted, and they all agreed that it was made up of twenty-one distinct joints. The antenna of a small moth was next examined, and found to have seventy joints. A part of the antenna of a very small moth, measuring only one-sixteenth part of an inch, was found to be composed of twenty joints. By a little calculation they found that each joint was only one three hundred and twentieth part of an inch in length, so that it takes three hundred and twenty of them to make an inch in length.

"How astonishing!" said Charlie. "Isn't it wonderful that such very small pieces can be put together so perfectly, and in such a way as to move on each other in all directions? I would like to know what kind of a joint holds these parts together. It cannot be like an elbow-joint, for then the antenna would have motion only in two directions. They must be jointed somewhat as our head is jointed to the neck, so that they can move in all directions."

"You are right," said Uncle Samuel, "in your last conjecture. The lower end of each part is rounded like a ball, and the upper end is hollowed out like a cup, and made to fit the lower end of the part above it. We call the joint the ball-and-socket joint, and as you see, it gives to each part motion in all directions."

Mary, as she was looking at the antenna of the *pieris* butterfly through the microscope, had her attention attracted by the little scales that covered

7

them, and with the enthusiasm of a discoverer ex-claimed :

" Just look here at the scales and spines that grow on the horns of this butterfly ! I should like to know what use they are to it."

" You are a bright little observer," said Uncle Samuel, " and your inquisitiveness will make you a philosopher yet. These scales are very small, in-deed, and in shape are similar to those found on the wings. My curiosity led me to make drawings of the very scales you have just discovered, which I will take from my portfolio.

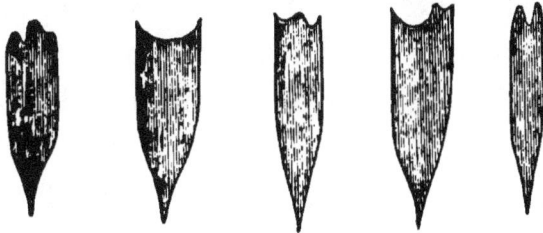

Fig. 26—Scales of Antennæ.

" Their color doubtless adds much to the beauty of this part of the butterfly's head-dress. Other-wise their use is not certainly known. Perhaps, if the antennæ are used as organs of hearing, the scales may serve to collect the waves of sound, and thus increase their intensity. Or if they are organs of touch, the increase of their surface by the scales may make their power of feeling more perfect. I cannot think that He who created the sun, moon and stars, and made man and so wonderfully en-dowed him, would make organs with so many joints

7

and scales and spines, and so entirely under the control of the insect, unless every part was designed to contribute to the continued life and comfort of the insect. I believe that He has fitted all His creatures for happiness in their respective conditions of being, and has given them all the functions necessary to the fulfillment of the design of their creation. So, although I cannot certainly tell you just how God meant that the insect should use all these head-appendages, there are some facts that help us to guess at their use.

" When a beetle is suddenly surprised by a sharp sound, it stretches its antennæ outward, as if in a listening attitude; and when the noise ceases, they are placed in their former position. Duncan tells us 'that, on close examination, a soft articulating membrane can be detected at the base of the antennæ, beneath which the antennal nerve is conducted,' and that this membrane may be considered as corresponding to the tympanum of the internal ear, and the nerve alluded to as the acoustic nerve. So the stock of the antenna, with its scales, may be regarded as analogous to the external ear, collecting the vibrations of the atmosphere and carrying them by means of the antennal nerve to the auditory organ at its base.

" Kirby says, ' How know we that every joint of some antennæ is not an acoustic organ, in a certain sense distinct from the rest? We see that the eyes of insects are usually compound and consist of numerous distinct lenses; why may not their external ears be also multiplied, so as to enable them with more certainty to collect those fine vibrations that we know reach their hearing organ, though they produce no effect upon our grosser organs?' I like the boldness as well as reasonableness of this conjecture. As God has multiplied the eyes of insects till they are counted by thousands, why may He not as profusely endow them with ears, all acting in concert and increasing their power of hearing?

" Here is a picture of a curious insect that lays its eggs in the bodies of grubs that bore in timber.

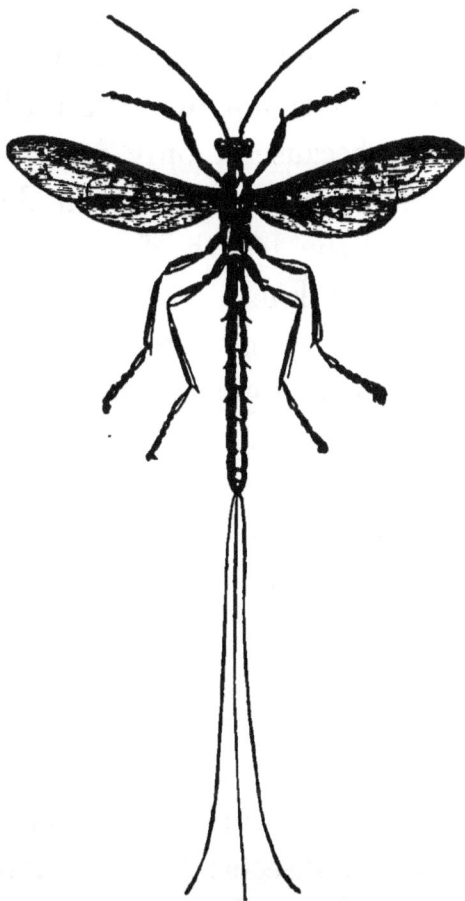

Fig. 27.—RHYSSA *lunator.*

It has three long tail-appendages, the middle one being a tube through which the eggs pass from the abdomen into the body of the grub. This is called the 'ovipositor, or *egg-placer.* It is here represented in the act of laying its eggs. Its name is *Rhyssa lunator.*

"Its antennæ are long and of great use to it. When it is searching for a grub in which to lay its eggs, it explores the hole made by the grub first with one antenna and then with the other, during which operation they quiver intensely to the very root. It is not the object of the insect to *feel* the grub, because it only inserts the antenna a short distance, and generally the depth of the grub is far beyond the length of the antenna. Its evident desire is to know whether there is any grub in the hole or not, and this it ascertains by the use of its antennæ as organs of hearing, catching by them the slightest sound produced by the gnawing of the worm.

"Bees when collecting honey first insert the tips of their antennæ into the cup of the flower. When the anthers are bursting or the nectar is exuding, a slight noise is produced, which is perceived by the inserted antenna.

"Kirby gives the following instance, which vividly illustrates the effect of sound on the antennæ of insects: 'A little moth was reposing upon my window. I made a quiet—not loud, but distinct—noise; the antenna nearest to me immediately moved toward me. I repeated the noise at least a dozen of times, and it was followed every time by the same motion of that organ, till at length the insect, being alarmed, became more agitated and violent in its motions.' 'In this instance,' continues he, 'it could not be *touch*, since the antenna was not applied to a surface, but directed toward the quarter from which the sound came, as if to listen.'

"The observations of some naturalists prove that, in addition to the function of hearing, antennæ are sometimes used as organs of *touch*. Huber tells us that when a beehive has lost its queen, the workers that first receive the intelligence quit their occupation and their immediate circle; and meeting their companions, they cross their antennæ, and slightly touching them make known the sad news speedily to the entire colony. As no one is disposed to doubt the accuracy of the observations of this eminent naturalist, the conviction presses itself upon us that, by a significant touch or stroke of the antennæ, bees hold conversation with each other. Ants examine with their antennæ everything they meet, and use them in holding intercourse with their companions. Figuier says that 'when an ant is hungry and does not wish to leave its work, it tells a foraging ant as it passes, by touching it with its antennæ; the latter approaches it immediately, and presents to it on the end of its tongue some juice it has disgorged for this purpose.'

"Still another function is conceived to belong to these organs. It is supposed that some wonderful property residing in them enables insects unerringly to foresee approaching changes in the weather; that is, the antennæ are to insects what the barometer is to the mariner. Before the coming of a storm, although the intelligence of man can perceive no

indication of its approach, bees that are foraging
return immediately to their hives. They seem to
see the gathering storm-cloud and to hear the thun-
der long before the lightning-flash, and like wise
men betake themselves to the shelter of their busy
homes. If a fine, pleasant day is to succeed the
present one, they will wander far from their homes
and remain away all day. If rain is approaching,
though all indications of wet weather are wanting
to us, they will venture but a short distance from
their hives and remain but a short time away.
Ants possess the same instinct; for though they
bring out their young daily to repose in the bright
sunshine, they are never overtaken by a storm.
This wonderful gift of foretelling atmospheric
changes, so common to many insects, is supposed
to reside in their antennæ. It may be that these
organs are peculiarly susceptible of the influence
of the electric fluid, and in this way communicate
to the insect a knowledge of the state of the atmosphere.

"While, therefore, we cannot tell with certainty
the various uses of the antennæ, it is quite evident
that they are designed for special ends which are
of great advantage to the insect. And does not
God, by the mystery in which these remarkable
organs are involved, design to increase our reverence for his skill and wisdom and teach us humil-
ity? We call these diminutive creatures insignificant, and some think we lower ourselves when

making their structure and history subjects of study. But small as they are, we find in them the hiding of the power of the great Creator as truly as in the mighty agencies that control the universe. How humble we should feel when we find ourselves searching in vain for knowledge which insects possess as a natural gift from God! ' How great are thy works, O Lord! In wisdom hast thou made them all.' "

Seldom has any lecturer as attentive listeners as had Uncle Samuel during his long talk. His young friends hung upon his lips with intense interest. Mary's inquisitiveness had called forth a satisfactory answer, and Charlie gave intelligent expression to his renewed convictions of the wisdom and goodness of God as manifest in the antennæ of insects. They could all now examine these organs in butterflies and moths with an interest they never felt before. As they stood before their uncle's collection of insects they were particularly impressed with the great difference, already alluded to, in the structure of the antennæ of the two classes of Lepidoptera.

Having examined for some time longer the antennæ of different tribes of insects, they separated for the night, committing themselves to the kind care of their heavenly Father.

CHAPTER VIII.

THE following evening, when the children entered their uncle's study, they brought with them a magnificent specimen of the *Papilio turnus*. It is called *turnus* after one of the ancient kings. It is one of the most splendid of butterflies; and, as it is swift-winged and hard to capture, it was valued by Uncle Samuel as a fine acquisition to his cabinet. Having his chloroform-jar near him, he placed his newly-captured specimen in it, and it was soon fast asleep. This was to them a new operation, and they asked for an explanation. Their uncle showed them that chloroformic air could not preserve life, and that in this way a butterfly could more readily and tenderly be prepared to take its place among other specimens in the

104

cabinet—that after it had been about twenty-four hours in the jar he would take it out and put it on the setting-board, on which he would place its wings and antennæ in the proper position, and keep it there till its muscles would become stiff and dry.

"When it was caught," said James, "it would stretch out and then roll up and put away under its lower lip a long trunk, just as if it did not know what to do with itself, and wanted to tell us that it didn't fancy being caught, and would like very much to pull itself out of our hands. I have often seen butterflies sitting on the edge of flowers with their trunks stretched out to their full length; and when ready to fly away, they would roll up their trunks and put them away out of the road. I should like to know what they do with this instrument. It must be of *some* use, for every butterfly I ever saw had one."

"I 'spect they eat with them," said Bertha.

"What a notion, Bertha!" said Mary. "Do you think a butterfly can *eat* with that long thing? Why, it isn't a mouth. The butterfly can't bite flowers with it. Maybe it holds itself up with it while it feeds on the flower. It is a curious thing, anyhow. I don't see how it can be of any great use, and I'm sure it's not so very handsome."

"You are very much mistaken in your philosophy, Mary," said Uncle Samuel. "If you should write a book on this subject and read it to a convention of insects, they would all stamp their feet and shout at your strange notions, and then, rising up all over the room, they would tell you that you were as ignorant as an owl, and that you had better come and live among insects and learn something before you attempted to teach others. Bertha is nearer right, if she did come from the city. It is an instrument that belongs to the butterfly's mouth, and is used for taking in its food; and if you should cut it off, it would die."

" Now, uncle," said Mary, who was a little skeptical, notwithstanding his keen criticism, " if it can *eat* with such a long trunk as that, I would like to know how."

" You know that all animals do not live on the same kind of food, so God has kindly given to each one such instruments as enable it to find and eat its appropriate food. The lion with its strong claws, its great teeth, and rough, file-like tongue, seizes its prey and devours it. The bills of birds are so formed that they can best procure and take in their food. The strong jaws of caterpillars enable them to cut and masticate the coarse leaf on which they feed."

" What does *masticate* mean?" asked Mary.

" I know," said Bertha. " I have heard my father and mother use it often. It means *to crush with the teeth, to chew.* I was thinking how a caterpillar could chew, when it had no teeth."

" That is a very sensible thought, Bertha. It is because the caterpillar does with its jaws what we

do with our teeth that it is said to chew or masticate its coarse food. The butterfly does not live on
animal food like the lion, nor on the leaves of trees
like the caterpillar, but on the honey that is found
away down in the cups of flowers. If it ever takes
its food like other animals, it must be provided with
a long tube-like tongue or trunk with which to
reach down into the cup that contains its food. So,
when the caterpillar is changed into the butterfly or
moth, its eating organs are changed to suit its new
tastes and new mode of life. When James saw the
butterflies stretching out their long trunks while
they were sitting on a flower or hovering over it,
they were taking their food. The trunk of the
butterfly is its eating organ. It is long, because
otherwise the butterfly could not reach its food. It
is hollow like a tube, so that the honey may ascend
it into the mouth of the insect. It is altogether
under the control of the butterfly; and when not in
use, it is put away in a groove prepared for it under
the lower lip. Could any instrument be better
adapted to the wants and habits of the butterfly
than such a spiral trunk? It is called also a *proboscis*—a word derived from two Greek words,
one meaning '*before*' (the eyes), and the other '*to
feed*,' and suggestive of the use to which it is
applied.

"I have here a microscopic view of a butterfly's
proboscis, which will help you to understand its
structure. (Fig. 28 represents the entire proboscis

partly unrolled and greatly enlarged. Fig. 29 represents the two divisions of the proboscis partly sep-

 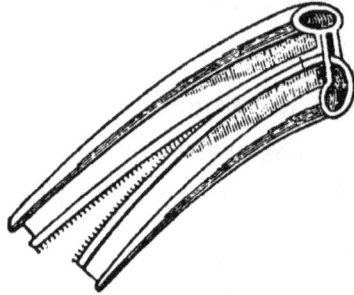

Fig. 28. *Fig.* 29.

arate. Fig. 30 represents a section of the proboscis, after Reaumur.)

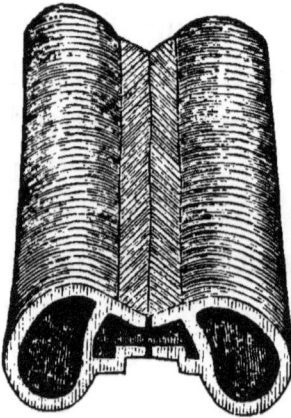

Fig. 30.

"This wonderful instrument is formed of cartilage, and composed of rings or transverse sections, as seen in Fig. 30. These sections are so put together as to be readily rolled up when not in use. Thus the Creator has not only provided this insect with a suitable instrument for taking its food, but so constructed it that without damage or pain it can be safely and promptly laid aside when not employed."

"How splendid," exclaimed Henry, "to have a place near one all the time into which to put away things when one is not using them! I think it would be a capital arrangement for James, for he is always losing his books, and he never knows where he lays down the axe or saw when he is done with it. One day he had to go to school without his slate because he forgot where he had put it, and he afterward found it in the wood-house —a queer place for a slate. He had better go to the butterfly and learn to be orderly."

"You are all guilty in this respect," said Uncle Samuel; "and I hope hereafter, when you have finished the reading of a book or the use of a tool, you will think of the butterfly, and go at once and put it in its place, that you may know where to find it when you want to use it again. By practicing according to this suggestion precious time will be saved, and valuable books and tools will be preserved.

"This proboscis is composed of two pieces, which can be separated throughout their whole extent. (See Fig. 29.) Each of these pieces contains a tube extending throughout its entire length; and being grooved on the inner side, they form, when united, another canal in the centre, of greater capacity than either of the other two. The junction is perfectly air-tight, and is effected by a countless number of fillets resembling those of a feather, which interlace and adhere to each other. The honey of the flower passes into the mouth through the middle tube. The use of the side tubes is not certainly known.

"Duncan suggests that they may be employed in transmitting air in aid of breathing. But why such an arrangement should be made for such a purpose, when the insect is so well provided with breathing-pores in its sides, I cannot see. I should rather believe the side tubes to be a mechanical provision by which the tube is strengthened and its flexibility increased. This suggestion agrees well with the general law observed in the structure of animals, by which bones are strengthened by being made tubular and their elasticity is increased. And such is the economy of the great Master-builder of the universe that He never uses any more material than is necessary for the accomplishment of the purpose designed in the structure of any organ."

"I see something like small feathers at the outer end of the proboscis," said Charlie, "and have been trying to find out of what use they can be. But I give it up."

"These filaments are membranous, and are not like feathers either in their structure or use. They are more like leaflets, and some naturalists think they are used to draw into the trunk the fluid on which the insect lives. Reaumur, after long and careful observation, regarded them only as so many points of support by which the organ is rendered more steady, and thus better adapted to the purpose for which it was designed.

"Here is a slide containing a butterfly's trunk. I will place it before the microscope and give Mary an opportunity of correcting her opinion about its beauty."

How surprised and astonished were our young naturalists as they one by one looked at the picture before them!

"Is that a *real* proboscis, uncle," asked Mary, "or are you just making believe?"

"How beautiful it is!" said Bertha. "Superb! I think the little butterfly would be *so* proud if it could only look at itself through the microscope."

"The colors are so richly combined," said Charlie, "one would think it was a perfect piece of mosaic. I don't wonder, Cousin Bertha, you think the butterfly might be proud to be able to flourish such an ornamental mouthpiece. Those barbarians I read about to-day, that wear great rings of ivory or gold in their noses and are proud of them, would be doubly proud if they had such a splendid ornament growing out of their under-lip. The peacock's tail don't compare with it in beauty, and yet you see how the lordly proprietor walks about our yard with his tail spread out and with stately step, just as if he owned the township. But I cannot clearly see why God should lavish the most brilliant colors upon things so small that their existence cannot be discovered without the use of the microscope."

"We cannot fully understand the works of God, but we do know that his wisdom is infinite, and that he is not at any time guilty of a wasteful squandering of ornament. He may love beauty for its own sake, and therefore clothe with richly-colored garments the smallest of his creatures. Or he may thus conceal the glory of his handiwork from the casual observer that its discovery may repay the diligent research of those who would seek for the gems of beauty which he has everywhere scattered among his works."

"How does the butterfly use its trunk when it wants its food?" asked James.

"Does it ever get its bright colors rubbed out?" asked Mary. "I don't think it would like to use it every day, for fear it might be soiled. Isn't it a beauty?"

"It knows how to take care of its beautiful dress," said Uncle Samuel, "as well as the most tidy little girl knows how to keep her books and dresses clean, so do not be concerned about the soiling of its trunk while it pumps up the honey from the centre of the flower. There is a beautiful description of the way the butterfly uses its proboscis in *The Insect World,* which one of you may read. I have marked it for your benefit."

Taking the book from his uncle, James read as follows:

"When it is fluttering round a flower, it will very soon settle on or quite close to it. It then brings its trunk forward entirely, or almost entirely, unrolled; very soon afterward it almost straightens it, directs it downward, and plunges it into the flower. Sometimes it draws it out a moment after, curves it, twists it a little, and sometimes even curls it partially up. Immediately it straightens it again, to plunge it a second time into the same flower. It repeats the same manœuvre seven or eight times, and then flies on to another." *

"The proboscis of flies" (Fig. 31), continued Uncle Samuel, "usually ends with two fleshy lips, and is provided with several fine bristles, which are sometimes as sharp as needles, and with these they pierce the skin of their victim. These bristles really take the place of jaws in biting insects, and hence people speak of being bitten by gnats or mosquitoes. (Fig. 32.) They often cause their saliva to flow freely into the

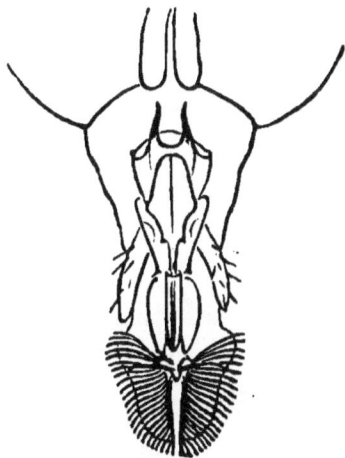

Fig. 31.—PROBOSCIS OF A FLY.

wounds they make, producing inflammation and itching. The saliva may prepare the blood which the insect extracts for more speedy digestion. Or Providence *may have* designed, by the pain inflicted, to warn us of their attacks, so that we may protect ourselves against them, and becoming their destroyers aid in setting bounds to their increase."

This conversation was kept up to a late hour, yet not one of the children became wearied, and Charlie, whose thoughtful mind suggested reflections sometimes beyond his years, said,

"I have learned several lessons from what you

* *Insect World*, by Louis Figuier, page 174.

8

have told us about this important and beautiful organ which God has given to the butterfly.

"One lesson is, that God is very good and kind to the butterfly. He has provided its food already prepared for its use in the cup of the flower, and has fitted up for it an apparatus by which it can reach that food and draw it up to its mouth. Will He not, therefore, provide for us? I think I can trust in Him more than I have ever done. I can now understand more clearly what you told us on the evening of the Lord's day, that 'His goodness is over all His works.'

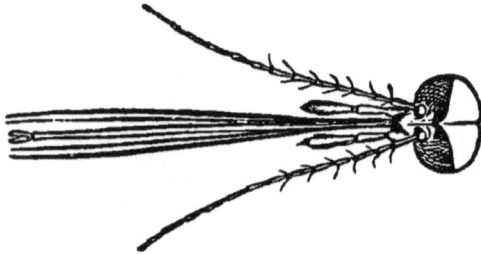

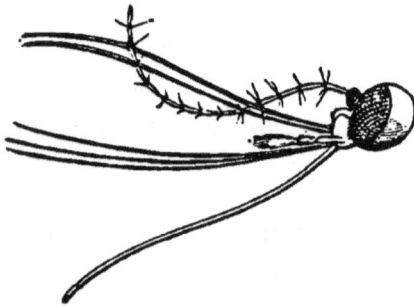

Fig. 32.—HEAD AND TRUNK OF FEMALE GNAT.

"Another lesson is, that if we get a living in this world we must work for it. The butterfly must fly off to the fields, hunt flowers, and use its trunk actively and industriously before it can get its breakfast. If it were idle and should stay at home and not work, it would soon starve to death, for it could receive no food from any of its sister-butterflies. So, if we are idle and refuse to work, we shall soon be without food. I recollect a verse in the Scriptures which says, 'If any would not work, neither should he eat.'* I see now that this is one of the laws of nature, and I will try never to forget it."

"Your lessons are very aptly drawn from the facts. But there is another lesson which they suggest. God has provided rich food in the gospel for hungry, perishing sinners. It is as free to them as the honey in the flowers is to the butterfly or the bee. As these insects receive no benefit from the honey laid up for them without going after it and drawing it up from the flower, so neither will you receive any benefit from the salvation which Jesus has provided for you unless you go to Him by faith, and receive it from His hands. Will you not all remember this lesson? and acting according to it, will you not go at once to Christ and receive the pardon of sins and eternal life? Jesus says, 'Whosoever will, let him come and take of the water of life freely.'"

* 2 Thess iii. 10.

CHAPTER IX.

IT was Saturday morning, and our young friends
were assembled in the family-room, according
to their custom on this day of the week, each
with Bible in hand, to study the Scripture lesson
for the Sabbath-school. Charlie opened his Bible
and read the lesson aloud. It was in the tenth
chapter of Exodus, and contained the history of
the plague of locusts, which God sent to punish
Pharaoh, king of Egypt. This event having taken
place about fifteen hundred years before Christ,
they wondered whether there were any such destructive locusts now.

116

"Don't you remember," said James, "how numerous the locusts were last year in the woods and orchards, and what lots of them we caught? I think they must have descended from the locusts of Egypt, for they kept all the time singing P-h-a-r-a-o-h, P-h-a-r-a-o-h. I can almost imagine I hear them now."

"Our locusts must be a very different kind from the locusts of Egypt, or else they have improved in their tastes and morals since the time of Moses," answered Henry. "The Bible tells us that 'they did eat every herb of the land, and all the fruit of the trees which the hail had left; and there remained not any green thing in the trees, or in the herbs of the field, through all the land of Egypt.' But the locusts of last year did not destroy the fruit nor eat a single leaf. They only injured the ends of the limbs by cutting a ring through the bark, so that they withered and fell to the ground before the eggs which were laid in them were hatched."

Charlie, who had observed the habits of these locusts more closely than his brothers, said, "I do not think that even this injury was as great as you represent it to be. The falling of the wounded limb was the exception rather than the rule, and I have observed that when the limb containing eggs fell off, the eggs seldom hatched. The great majority of the wounded limbs remained green and recovered. The eggs in these limbs grew large before hatching, while the eggs in the dry and dead limbs were always shriveled. I noticed that between the 20th of July and the 1st of August the eggs that remained on the trees hatched, and that the young locust dropped to the ground of its own accord. The fall never hurt the little thing, because it was so light that it fell as gently and as softly as a feather. It seems to me that the life of the limb is necessary to the health of the egg, and that the circulation of the sap in the branch somehow helps the egg to hatch out the young. I don't understand it, but I believe what I see, and I know there must be some relation between the life of the limb and the size and hatching of the egg."

Mary's inquisitiveness was excited by the story of the young locust falling to the ground, and she asked, " How did the little child of the locust know that it had to get to the ground? and who told it that the ground was below it, and that it could get there by letting go and falling?"

" It is very strange," answered Charlie, " how the young locust always knows so well how to act and where to go. It seems to me that insects become very wise just as soon as they are born. They always know just what to do, and how to do it. They do not need any mother to teach them, nor do they stop to reason, or even to think, before they act. The little locust not a day old falls to the ground just as if it had done so a hundred times, and as soon as it touches the earth it begins to burrow its way down to the roots of the plants or trees from which it is to draw the juices that nourish it. God gave the young insect this knowledge, and taught it how to find its own food. We ought to love God, who is so kind to such apparently useless little creatures, and believe that he will take still better care of us if we trust in him and serve him."

Just then the door opened, and Uncle Samuel entered and took his seat in the cozy arm-chair that was always ready for him in the family-room. It was a pleasant scene that greeted him. Around the table were seated his three nephews, with their little sister and their city cousin, having their open Bibles before them. They were evidently intensely interested in the study of their Scripture lesson; and when their uncle entered, their countenances were all aglow with delight, for they knew that he would help them out of their difficulties and answer any questions they would ask him.

" I am glad to see you so diligent this morning," said their uncle. " This is your Bible morning, I perceive, and you seem to be studying like little theologians. What is the story that has interested you so greatly?"

" It's the story of the plague of the locusts," answered Henry, " and we have been talking about the locusts that were so numerous in our region last year. James thinks that they came all the way from Egypt, but I don't think they did. Now, we want to know which is right. Did last year's

locusts descend from those that ate up every herb
in the land of Egypt?"

"No. They are not locusts at all. People mis-
name them *locusts* because they come in such great
numbers, and so they imagine that they are always
crying P-h-a-r-a-o-h. But they ought not to be so
called, since they belong to a very different order of
insects.

"The 'locusts' you saw last year belong to the
order of insects called *bugs*. This order is divided
into two groups. One of them is known as the *true*
bugs, and are such as the *squash*-bug and the *plant*-
bug. The other contains the harvest-flies, the plant-
lice and bark-lice. The wings of the latter are thin
and clear throughout. They have wing-covers, but
they are not dark and thick, like those of beetles,
but transparent, like the true wings. Such wings
you doubtless noticed on the locusts, for they belong to this group, and are a species of harvest-
fly.

"Scientific men call this order of insects, from
the kind of wings they have, *Hemi*ptera—that is,
half-winged—and the first group they name the *true
Hemiptera*. The second group, to which our locust
belongs, they call Hemiptera *homoptera*—that is, *like*-
winged Hemiptera—because the wing-covers are of
the same texture throughout.

"There is another order of insects, called *Ortho*-
ptera, or *straight*-winged, because they all fold their
wings lengthwise, like the folding of a fan. It is

to this order that the true locust belongs. It cannot, therefore, be the ancestor of the locust that infested our region last year, any more than the lion could be the ancestor of the cow.

"The correct family name of *our* locust is *Cicada* —the name which the ancient Latins gave to *harvest*-flies. The chief peculiarity of this harvest-fly is that its larva lives under ground seventeen years before it is ready to change into its winged or perfect state. From this fact is derived its specific name of *septemdecim,* which is the Latin word for *seventeen.* We have, then, its true name in English —the *seventeen*-year harvest-fly. There is also a so-called *thirteen*-year locust, called *Cicada tredecim,* although it has not been yet decided that it differs specifically from the seventeen-year cicada. It is not, then, as you see, even a distant relation of the Egyptian locust, and it ought to bring a suit against its defamers for slander when they persist in calling it a *locust.*

"You all know that this so-called locust did not eat the leaves of trees. It has no taste for that kind of food, and no mouth-apparatus for cutting and chewing the leaf. It is provided with a piercer and sucker, by which it penetrates the tender bark of the trees and drinks the sap. Of this kind of food it is very fond, and cannot live on any other. You might as well expect a house-fly to eat the leaves of the trees as our locusts to do so.

"I have some drawings of this insect in my portfolio which will greatly assist you in the study of its history, if one of you will please bring it down from my study."

This was very quickly done by Bertha and Mary, who had become so much interested in their uncle's remarks that they were very anxious to see the illustrations.

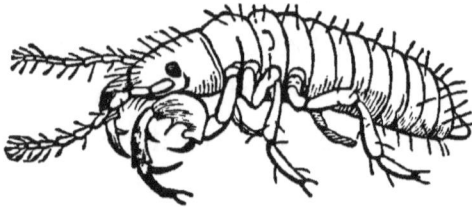

Fig. 33.

" Here is a greatly-magnified picture of the larva

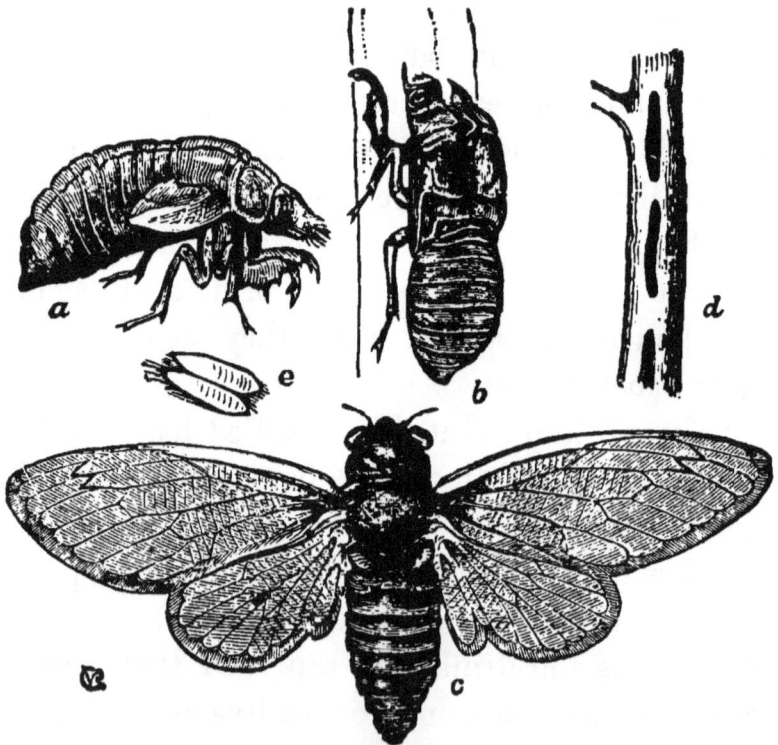

Fig. 34.

(Fig. 33) just as it comes from the egg. It is very small when hatched, but is well supplied with large and powerful front legs armed with strong claws, with which to dig its way through the world. Look, too, at its long eight-jointed antennæ; and if it was the living larva, you would see how rapidly and skillfully it uses them.

"In this group (Fig. 34) you see this insect in the other three stages of its life; *a* represents the *pupa; b,* the pupa-skin after the perfect insect has escaped; *c,* the perfect insect; *d,* incisions in bark of limb for receiving the eggs, *e.*

"When the cicada-mother is ready to lay her eggs, she clasps the selected branch with her legs, and then, bending down her piercer (Fig. 35), she thrusts it into the bark and wood so as to bore a hole obliquely to the pith. By repetitions of the same operation she makes a fissure of sufficient length to receive from ten to twenty eggs. She now proceeds to fill the cavity with eggs by means of the piercer, which serves as an egg-layer, or, as scientific men say, an *ovipositor.* When she has done this, she removes a little distance and makes another nest in the same way, filling it with eggs. It takes

Fig. 35.

her about fifteen minutes to make a fissure and fill it with eggs, and she not unfrequently will make fifteen or twenty fissures in the same limb. The punctured twig is represented in Fig. 34, *d.* Some of these twigs die and fall to the ground, while many others live, and when healed of their wounds present the appearance of Fig. 36. The beak or mouthpiece of the cicada, by which it takes its nourishment, is represented in Fig. 35. This is also the instrument by which it stings, and not the ovipositor.

"To protect themselves against the water in low and wet places, the pupæ, when about to transform into perfect insects, continue the galleries which they ordinarily make to the surface of the ground, as represented in Fig. 37, *a*, full view, and *b*, sectional view. In the upper end of these chambers the pupæ wait quietly the time of their change, and then back down to below the level of

Fig. 36. the earth, as at *d*, and issuing forth from the opening *e* attach themselves to the nearest object and pass through the process of freeing themselves from the pupal case, leaving the dry empty shell for the joyous freedom of their new life.

"This structure is a remarkable instance of the provision which insects make to protect themselves against injury from natural forces which must be foreseen in time to make the provision. Let it excite our admiration of the wisdom of the wonder-working God, who alone could endow this insect with skill and give it instinctive knowledge of the very time when this skill is to be used."

"Uncle," asked James, "if the locust that we

saw last year is not a *true* locust, what kind of an insect is the Colorado grasshopper, that did so much damage in the West last harvest? I should think they ate leaves with a vengeance, for there was not

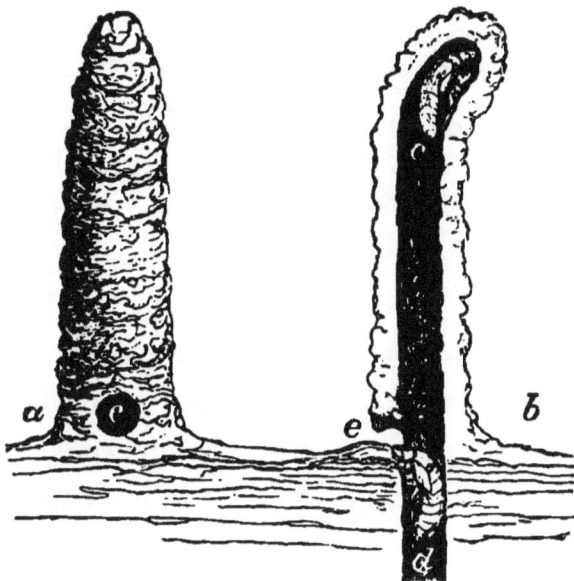

Fig. 37.

a leaf left on the trees nor a spear of grass in the fields on which they settled."

"This grasshopper, called *Caloptenus spretus,* or the *hateful grasshopper,* has become very destructive to vegetation in our States west of the Mississippi of late years. It belongs to the order *Orthoptera,* and is doubtless a member of the same family to which the Egyptian locust belongs.

"It does not differ much from the common red-legged grasshopper, with which you are familiar, as these cuts will show. Fig. 38, *a,* represents the *C. spretus;* b the *C. femur-rubrum,* or red-

legged grasshopper.   You notice that the wings of
the former are about a fifth longer than the wings
of the latter, by which it is enabled to fly miles at

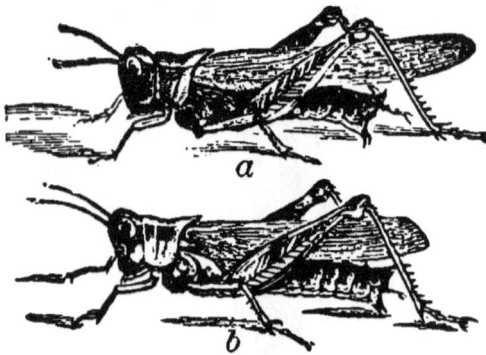

Fig. 38.

a stretch, while our
common species can
only fly a rod or two
at a time. The 'hate-
ful grasshoppers' fly
ordinarily from five
to ten miles in a
day ; and when on
the wing, they make
a rushing and roaring noise, hide the sun from view,
and appear like so many moving snow-flakes. When
they light down, 'they soon make a clean sweep of
every green thing, occupying and possessing the whole
country as they slowly proceed from point to point.'

"The locust that was sent to plague the Egypt-
ians was a kind of grasshopper, and, like the hateful grasshopper, was provided with jaws for cutting
leaves and grass and gnawing the bark of trees, and
with organs just suited for the digestion of such
heavy, coarse food. It could no more live on the
juices of plants, as the harvest-fly does, than the
tiger can live on grass."

"Have you a specimen of the true locust in your
cabinet, uncle?" asked Henry.

"I have not; but I have here a picture of a species of locust very abundant in the East, called the
migratory locust, or, as the scientific books have it,

Locusta migratoria. (Fig. 39.) This locust is believed to be the lineal descendant of the locusts that
are described in the Scriptures.

"Since the time of Moses a great many genera-
tions of the migratory locust have come and gone.
In the mean time the world has grown wiser and
better, but nothing has improved the tastes, manners or morals of these destructive pests. They are
as unchangeable as the 'infallible pope.' They
keep up to this day the daring and ferocity of the

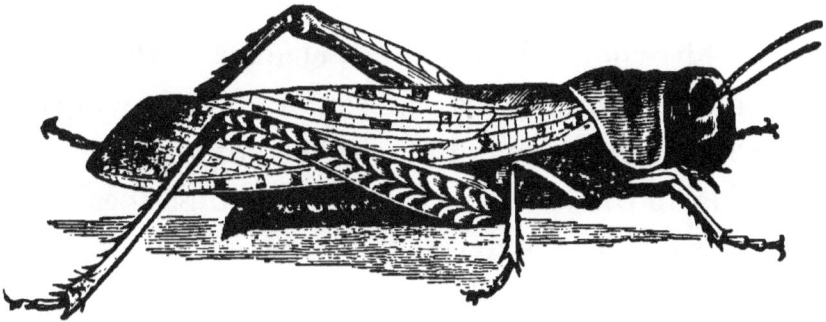

Fig. 39.—Locusta migratoria.

days of their ancient crusades, and settling in innumerable hordes upon the woods and cultivated
fields of Eastern countries, they leave 'not any green
thing in the trees nor in the herbs of the field.'"

"Look!" said Bertha; "it is just like a great big
grasshopper. What a large head, and what strong
legs! Wouldn't I be afraid if one of them would
light on me! I suppose all the little Egyptian
girls would run away from them when they saw
them coming."

"Yes, Bertha, it resembles the grasshopper very

much, only its horns are proportionately shorter. The thighs of its hind legs are thicker, so that it can leap much farther than the common grasshopper. It has also a greater power of flight, for the wing-covers, being narrow, do not prevent its free passage through the air, while its large wings, forming half a circle when expanded, are moved by very strong muscles, by which its flight is sustained for hours.

"When a large number of them take flight at the same time, they make a sound like the rushing of a whirlwind. In the ninth chapter of the Revelation, and ninth verse, the sound of their wings is described 'as the sound of chariots, of many horses running in battle.' The prophet Joel describes their sound as 'like the noise of chariots on the tops of mountains, and like the noise of a flame of fire that devoureth the stubble.'

"I must not forget to tell you of a singular arrangement which the female has for the purpose of burrowing a hole in the ground in which to deposit her eggs. At the end of her body are four short wedge-like instruments, placed in pairs above and below, so as to open and shut opposite to each other, making a double pair of nippers, with four short blades instead of two.

"Now let us imagine one of these insects about to lay her eggs. See how carefully she selects a spot of ground suited to her operations. Fixing herself in the proper attitude for her work, she drives her little wedges into the earth ; and opening and withdrawing them, she enlarges the cavity. She then inserts them again and again, repeating the same operation each time, until she has made a cell about an inch and a half deep. This cell and its tubular entrance she coats with a kind of glutinous matter. She is now ready to deposit her eggs, which are usually about seventy in number. She does not, like some insect-mothers, lay in a store of provisions for her children, for she knows they will look out for themselves as soon as they escape from the egg. Wood tells us that ' the young do not attain their wings for three years, and during that period are called in Southern Africa by the popular and expressive Dutch name of *voet-gangers,* or *foot-goers.*'

"They have no voices for song, but the males have a remarkable taste for instrumental music. They carry their little fiddles with them, and wherever they alight they keep up a constant chorus of strange and noisy music. The veins of the under side of their wing-covers are the strings of their violins, and under them, in their body, is a deep cavity with a thin piece of skin stretched tightly over it, like a drum-head. This is the body of the violin, which increases the sound. Their hind legs are the bows which play on these veins or fiddle-strings, and make them vibrate. When a locust begins to play, he bends his hind leg under the thigh, where it enters a furrow made for it; then

9

he draws the leg swiftly across the veins of the wing-cover. He does not play both fiddles together, but alternately, first upon one, and then on the other. This music is so pleasing to the ear of Spaniards that it is said that they often keep them in cages at their homes for the sake of the tunes they play on their fiddles.

"Their great appetite makes it necessary for them to go from place to place in search of food; so God has given them great and strong wings, and although 'they have no king, they go forth, all of them, by bands.' In the evening they gather together in one place and rest upon shrubs and trees, sometimes breaking the limbs with the weight of their numbers. In the morning, as soon as the sun has made it sufficiently warm, they take wing again, but not without leaving behind them the earth stripped of its greenness and the shrubs peeled of their bark. Desolation marks the course of their journeyings.

"But they are not without their enemies. They furnish food to beasts of prey, birds, serpents, lizards, frogs and men. Moffat tells us that the people of Southern Africa consider them good food, and capture them by the thousands. They prepare them for eating by boiling or steaming them. After boiling a short time they are spread on mats in the sun to dry; then they are winnowed to clear away the wings and legs; after which they are put in sacks or laid in a heap on the floor of the house, to be used as needed. They eat them whole or make meal of them; they prepare a kind of stir-about, of which they are very fond, and on which they become very fat."

"Is it in Africa alone that these locusts are found?" asked James.

"They are found also in Italy, in the south of France, in India and in China. Everywhere they are regarded as a great curse, and very curious means have been used to banish them. The negroes of Soudan try to frighten them by savage and hideous yells. But, regardless of their yells, they go on with their work of destruction till they can find nothing more to destroy, and then they leave of their own accord. In Hungary the people fire cannons at them. But they are as fearless of gunpowder as of savage yells, and do not mind the roar of the heaviest artillery. In the south of France the government pays a royalty for the destruction of their eggs, which is a more sensible and sure way of diminishing their number.

"In the Middle Ages, when the priests of Rome had everything their own way, these infatuated men thought they could drive them out of any country by heaping the curses of the pope on them. I once read a story about a monk of Ethiopia who undertook in this way to rid the country of these pests. At a time when they were very numerous he assembled the natives and told them about the great power of the pope—how he controlled everything

in heaven above and in the earth beneath—and that
in his name he would command the locusts to leave
the country, and they would promptly obey him.
He then asked them to chant some psalms with
him, and after this act of worship he addressed
himself authoritatively to the greater congregation
of locusts, which hung over them in a dense cloud,
shutting out the light of the sun. He told them
how wicked it was to destroy the gardens of the
Christians, and thus to sin against the Church and
the pope, and excommunicated them—that is, put
them out of the Church, giving them over to Satan.
He then charged them, as they valued their future
welfare, to leave immediately the land of the Christians and go either to the sea or to the land of the
heretical Moors."

"And did they go away?" asked Mary, who was
beginning to feel her sympathy awakening for the
poor locusts, and to wish that they would save
themselves from the curse of the monk.

"Not at all. The self-willed locusts never let
on that they heard him, but settled right down
on the trees and shrubs and plants, and began their
work of destruction, having no respect for the gardens or orchards of Christians."

"And wouldn't the old monk be very angry at
them for not doing what he told them?" asked
Bertha.

"I suppose he was, for he seized a number of
them, so that he might be certain that they at least

heard what he had to say, and, shouting aloud, he called on all the birds and animals and tempests to drive these pests away; and, letting his captives go, he told them to tell their companions what terrible punishment would be inflicted on them by Heaven if they did not leave the country immediately. But they were as obstinate as ever, and never thought of leaving till they had destroyed every green thing."

"What a fool that monk must have been!" said James. "He ought to have known that the locusts could not understand a word he was saying, and that they did not belong to the Church and would not care for his curses."

"It shows," said Charlie, "how much ignorance there was in those days, and how dark were the minds of even the wisest of churchmen. We ought to be very thankful that we have the Bible to teach us the truth and are not misled by the popish faith of the Catholics. I have no doubt the monk was a wiser man after his experiment, and that he had less confidence in the power of the pope over such obstinate heretics as the locusts."

"The Arabs," continued Uncle Samuel, "believe that words of prayer addressed to Mohammed, written on a piece of paper and put in a hollow reed planted in the middle of a field covered with locusts, have power to compel them to leave. Sometimes they take four locusts, and, writing on the wings of each a verse of the Koran—the Mohammedan Bible—they let them fly into the midst of the swarm, and the destructive army, they say, leaves for other parts. I have no doubt, however, that the locusts care just as little for the Arabs' prayers and the Koran's verses as they do for the curses of the popish monk."

" Why are they called locusts?" asked Mary.

"If you look into your dictionary, you will find that the word *locust* is made up of two Latin words which, put together, mean *burnt place.*"

" Oh, now I see," said Mary. " It is because the ground, after the locusts leave it, is so stripped of everything green that it looks as if it had been burnt over with fire."

" That makes me think," said Charlie, " how very careful we should always be to do that which is honorable and right. The locust gets its name from its work. Its work is bad, and its character and name are bad also. It is not because it is an *ugly* insect that men hate it. I think it's a noble-looking little animal, with its large head and great wings, and if it only had a good name it would be very much admired and its visits would be welcomed. But it has brought itself into disrepute by what it does. Its consequent bad name follows it wherever it goes. Now, I think we ought to take a lesson from the experience of the locust. If we do what is wrong when we are young, and persist in it, we shall get a bad name, which will follow us wherever we go; and sometimes this bad name fixes itself on boys so that they do not get clear of it all their lives.

" There is one of our neighbors who came here from a great distance, and although no one knew anything about his previous history, and he was at first esteemed by everybody, it was not long till some one came from the place where he was raised and told us that when he was a boy he was known to be very stingy, and they called him *Stingy* Joe, and that he was so called when he moved away; and now everybody calls him *Stingy* Joe, and his son is called *Stingy* John and his daughter *Stingy* Jane. How unfortunate it is for him and his children that he was so illiberal when a boy! for, though he is as liberal now as other people, he cannot get clear of his bad name."

" I hope you will follow the suggestions of Charlie," said Uncle Samuel, "and remember the lesson which the history of the locust teaches. Be true to yourselves. Solomon said a long time ago that ' a good name was rather to be chosen than great riches.' Some men do not believe what Solomon said, and sacrifice their good name for the sake of riches. Do not follow their example. Begin your life in the fear of the Lord. Live so as to secure his favor, and you will always have the favor and love of your fellow-men. Good people will not then shun you or seek to banish you from their society. Your company will be sought and prized by all whose acquaintance is worth having, and you

will leave a *good name* as an inheritance to those
who shall come after you."

Just here this long talk about locusts was interrupted by the sound of the dinner-bell, and soon
the happy household were discussing matters of
general interest around a plentiful table.

EXODUS. X. 4.

CHAPTER X.

THE first day of May had come and gone, the
tender grass-blades had covered the lawns with
their clothing of velvet-green, the flowers of spring
were opening their delicate petals to the sunlight,
the wild birds were caroling the songs of the last of
the spring months, and all nature was smiling with
the life-beauty of the virgin year, when Charlie
and James and Henry proposed to accompany their
uncle in one of his insect-rambles in the meadow
not far from their cheerful home.

Mary, whose thirst for knowledge was not a whit
behind that of her brothers, on her own behalf and
that of her cousin Bertha asked permission to ac-

company them. The request was readily granted,
for all felt that the party would be incomplete without their sprightly, thoughtful cousin and their inquisitive little sister.

Uncle Samuel was always glad to have his young
friends with him when he visited the homes of insects and studied their domestic habits. It added
very much to the attractiveness of the walk to witness the cheerfulness and eagerness with which they
searched for rare specimens and captured them.
Their questions too, often so original and intelligent, compelled him to be always ready to tell
them the interesting facts with which the life and
habits of insects abound. These familiar talks had
inspired them with intense love for the study of
insects, and by the training thus obtained their
minds were rapidly developing. They were learning to think and to reflect, and were already almost
worthy to be called young naturalists. Nor did
they confine their study to insects, for their uncle
taught them a great many facts about plants and
flowers, and also about birds and animals; and
when they would find a curious stone he would
tell them its scientific name, how it was formed,
and the age of the world to which it belonged.
By communing with Nature they had learned so
many interesting facts which could not be found
in their school-books that they were always eager
to learn the lessons which were taught by the wonderful things that God has made.

They had not been long in the meadow before
their attention was called to a large bee with a short,
robust body, having bands of very bright colors
around it. It made a humming noise when it flew,
and its hind legs were armed with two great spurs,
as if it belonged to the knights of whom we read in
history. Our young naturalists had never before
seen such a bee, and they were very anxious to learn
its name and history; and especially so because they
discovered that it had lighted upon a spot in the
meadow where there was a little hollow made, into
which it entered and immediately began to increase
the excavation. In this way it showed that it had
been at work in that place before, and it was now
carrying on to completion some plan already
formed.

Patient naturalists are always willing to remain
near an insect and learn from its movements what
it purposes to do. It is in this way that all our
knowledge of insect habits and instincts is obtained.
But our young students could not wait on such a
long and tedious process to learn what this busy in-
sect-laborer was doing. Their restless natures would
not permit them to go to the bee itself for the in-
formation which they wanted. So, gathering around
their uncle quietly, lest they might disturb the industrious insect, they asked him to tell them its name
and its history, and why it was working so earnestly in the ground.

"Is it going to bury itself," asked little Miss Inquisitive, " as the great green tomato-worm of which you told us does?"

"Why, it has wings already, and it is *so* beautiful," quickly responded little Thoughtful. "Surely it does not need to become a pupa, and then change to something else."

"This very industrious insect introduces you to a new order of insects, called *Hymenoptera*, from two Greek words meaning *membrane-winged*. You can see how clear the wings of this bee are, and how few nervures they have. This peculiarity is common to all insects of this order. They have four wings, which lie horizontally upon the body when not in use. Their mouth is provided with jaws and lips adapted for suction.

"In this order we find the most industrious and intelligent insects. They build their houses with great skill, using them as nurseries and as storehouses for their provisions. They are very careful of their young and watch over them with true maternal affection. Many of them form republics or monarchies of their own, and govern them with laws that are rigidly enforced.

"Its four most remarkable families are the bees, the wasps, the ants and the gall-insects. The insect that is working so diligently before us belongs to the family of *bees*, called *Apidæ* by scientific men. There are several *kinds* of bees. The kind or *genus* to which this bee belongs is the *humble-bee.*

" It is commonly called a *bumble-bee.* The common name of every species of humble-bees is *bombus*—a Latin word meaning *humming* or *buzzing*, and descriptive of the peculiar sound of its wings when flying. This bee is known as the *moss* humble-bee, or the *carder*-bee, and its scientific name is *Bombus muscorum*—the latter word meaning *moss.* You observe that it is not as large as the common humble-bee, but rather shorter and thicker than the hive-bee. It is now engaged in preparing the foundation of a house for itself and family.

" In the community of humble-bees there are three classes—males, females and workers. During the long winter all the males and the workers die, and the females alone survive. These pass the winter under ground, in apartments separate from the nest, and which each one fits up for herself with a warm carpeting of moss and grass. In the early spring, when the sun warms their winter retreat, they make their appearance to lay the foundations of new colonies and to build up new communities of the bombus family. In this enterprise each bee works for herself, and elects herself the queen of the new city which she means to build and people with a small but busy multitude.

" This bee is just now laying the foundations of an empire which is to be indebted to her alone for its existence. Her little city when completed will not exceed six or eight inches in diameter. Its dwellings are to be rudely built, and its streets to be arranged without regard to order or refined taste. Its inhabitants may not exceed twenty, while they may number two or three hundred.

" When the mother-bee has completed her excavation, and advanced far enough with her work to afford protection to her offspring, she constructs a few cells, and puts in them a paste made of pollen and honey. She then lays six or seven eggs in each, and closes them in. The little grubs that come from the eggs live together peaceably, as brothers should, eating at the same table and sleeping in the same chamber. At first the cell is only the size of a pea, which in a short time becomes too small to contain the rapidly growing grubs, and splits in several places. The mother-bee, however, soon repairs the walls of her cells by filling the cracks with wax. In this way these infant homes are made to increase in size, so as to give sufficient room to the growing inmates.

" When the little worm, the baby-bee, is ready to pass into the pupa state, it ceases to eat, and spins for itself a cocoon of very fine white silk. Inside of this soft shroud of silk it remains about fifteen days, during which all those wonderful changes are going on that prepare the bee for the great work of its life. When it is ready to appear in its winged state it finds itself unable to work its way out of its silken case. It has shut itself up, and it must now perish unless it is helped by its mother or brother bees. Their instinct comes to its relief.

In some wonderful way—perhaps by a significant motion or noise made by the helpless inmate of the cell—they know the precise time when the perfect insect is developed, and they gnaw off the covering as ants do in the same circumstances, and the released prisoner leaves the dark abode of its early life to enjoy the pleasures of its citizen life in the rising city.

" The first bees thus added to the community are workers, and they immediately employ themselves in building new cells and in raising and fortifying the walls of the bee-city. These walls are built of wax, and, like a rampart, rise up at all parts of the circumference. The whole is covered with a dome of moss, or of withered grass if moss cannot be had, closely packed or woven together and lined with wax. The entrance is in the lower part, and leads through a gallery or covered way about a foot long and half an inch in diameter."

During all this time our young philosophers were observing closely the operations of the bee, and listening with intense interest to their uncle's story. At length, Henry, whose imagination had been quickened by his uncle's description of the humble-bee's nest, said,

" How beautiful their little city must be, with its wax walls and wax houses, its covered halls and its streets! Wouldn't it be nice if we could visit them and see how they all live in their moss-covered city ?"

"You forget," said Uncle Samuel, "that I told you that their houses were not built with very much taste, and that their streets were very irregular. These bees are not so refined and neat in their style of living as their cousins, the hive-bee and the wasp.

"As I have some drawing-paper with me, I will gratify your curiosity ·by drawing a few pictures representing their nest. I will first make a drawing of the cells as they appear inside of the nest. (Fig. 40.)

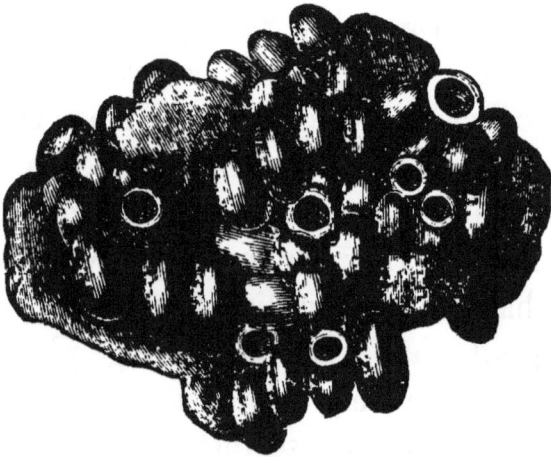

Fig. 40.—CELLS FROM A MOSS HUMBLE-BEE'S NEST.

"These are their breeding-cells. The wax of which they are made is brown-colored, and the cells are placed without much regard to order, and in shape they are somewhat flat and gobular.

"And now I will give you a representation of the entire structure, with the opening of its covered entrance into the interior (Fig. 41), and of the bees when they are carding or matting the moss for use. (Fig. 42.)

"If you could examine the interior of one of these wonderful homes, you would see in all the corners and in the middle of the combs a number of vessels shaped like a goblet filled with honey and pollen. These vessels are the empty cocoons left by the larvæ, and the honey and pollen stored

Fig. 41.—Moss Humble-Bee's Nest.

in them are the provisions which the bees have laid up to supply their daily wants. They display wise economy in thus making use of the cocoons, which would otherwise have been so much rubbish in their way."

"What a happy time they must have!" said Bertha. "Wouldn't it be nice to be a moss humble-bee,

10

and live in such a cozy nest and eat honey all the time ?"

"They do not always live so very happily," answered her uncle. "The workers are very fond of the eggs laid by the mother-bee. If not prevented,

*Fig. 42.—*Moss Humble-Bees carding Moss.

they drag them from the cells and devour them; so the mother is compelled to defend her eggs, and many a conflict takes place between her and the workers; but by constant watchfulness and some vigorous fighting she generally succeeds in driving off the thieving bees. There is a number of females which lay eggs of workers only. Of this class the queen-mother is very jealous, and she often drives them away from their cells and eats their eggs. So you see that selfishness reigns in this little community, disturbing its peace just as it disturbs the peace among men, who are in this respect no better than insects.

"But you must not think that the workers are always trying to destroy the eggs. They are only fond of fresh eggs; and if the mother keeps them off for a short time, they lose their taste for them and take their part in rearing the young grub, and are as affectionate and careful as the mother.

"When the temperature of the nest is too low for the hatching of the eggs, the males and females sit upon them just like a hen, and thus increase the warmth of the eggs. If in any way the comb becomes displaced so as not to stand firmly when the bees are sitting on the eggs, two or three other bees will get on the edge of it, stretch themselves over it, and with their heads downward fix their fore feet on the ground, and with their hind feet keep the comb from falling. In this way they will continue to hold firmly the tottering comb till pillars of wax, built by other bees, are put under it to support it."

"How do the bees know so well just how to prop up a tottering comb, if they act only by blind instinct?" asked Charlie, who always wanted a true philosophical reason for everything.

"They do not act as mere machines when they deliberate and plan in such a manner. The instinct of insects is not always *blind*. It gives as clear evidence of being guided by reason in some cases as the mind of man does; and this fact should increase our admiration of the wisdom and beneficence of God manifested in their intelligent actions."

"I have been wondering," said Bertha, "how they gather the moss and mat it together and put it on the roof of their house. I think they must have a *good deal* of knowledge to find the moss and weave it together close so as to make such a nice roof."

" This is one of the most curious facts in their wonderful history. One would suppose that each bee would cut off with its mandibles a single blade of moss or of withered grass, and then fly to the nest and lay it in its place, as the dauber-wasp carries each pellet of mud and puts it in its place in its cell; but it does not. When the mother-bee is alone, as is usually the case in the spring, she pushes little bundles of moss or grass along the ground, walking backward, till she gets them to her newly-built home. After other bees are hatched out, a number of the workers go out moss-gathering together, and arrange themselves in a line reaching from the nest to the moss-bed, the head of each one being turned away from the nest and toward the moss. When they are prepared to say 'All ready!' the first bee in the line lays hold of some of the moss, and having *carded* it with her fore legs into a sort of felt, she pushes it under her body and throws it with a kick to the next bee, and she passes it in the same way to the next, and so on till it reaches the nest. When the materials are all collected, the workers proceed to manufacture the cover that rests like a dome on the nest. When completed its height is from four to six inches above the level of the ground. Underneath the moss or grass the vault is ceiled with coarse wax to keep out the rain and to protect it from high winds. You have here an opportunity of making observations for yourselves on the habits and instincts of this interesting member of the bee family which I hope you will not neglect. It will repay you well to visit this growing city frequently and note its progress; and you will be able to verify what I have told you, and you may perhaps be rewarded by the discovery of some facts heretofore unknown about the domestic habits of these industrious insects."

"Wouldn't that be nice?" said the impulsive Mary, who seemed to be feeling already the joy of some actual discovery.

"I have just been thinking," said Charlie, "what a capital teacher this humble-bee is. Even if it is *humble* we ought to listen to it. It doesn't speak in words, but it does in acts. You see it is ambitious to do some good thing while it lives; and although it has no one to encourage it or help it, it works earnestly and cheerfully. Everything it does helps it to gain the object of its ambition. Doesn't that teach us that ambition to do right and to do good ought to make *us* work earnestly and cheerfully? Ought we not to resolve to accomplish something great and good while we live, and then live for that end? If we do so, we shall be as likely to succeed as the humble-bee. I shall come here often, just to learn from this chubby insect how I ought to live."

"You are quite a philosopher," said Uncle Samuel. "I hope you will carry out your purposes; and if this morning's ramble inspires you all with better and nobler views of life, I shall not regret the time I have spent in telling you the story of the moss humble-bee."

"And I shall never see a humble-bee without remembering the lesson of this pleasant May day," said Mary.

CHAPTER XI.

A FEW days after the meadow ramble which we have just described, Bertha and Mary visited the nest of the moss humble-bee to observe what progress the mother-bee had been making in the mean time. They found her moving rapidly from place to place, never idling a single moment, and that she had already commenced making her wax cells. They became so much interested in her cunning devices that they were unconscious of the rapidity with which time was passing. At length,

fearing lest their long absence might excite alarm,
they started to the house. A short distance from
the humble-bee nest Mary observed a small mound
of fresh earth which seemed to be the work of some
insect. This immediately attracted their attention.
Just at the top of the mound they found an opening
about the diameter of a pipe-stem. While gazing
into the opening with lively curiosity, and wondering what insect could have built such a mound, and
where it got the little fine pellets of black earth of
which it was composed, a bee about one-third of an
inch long flew out of the opening. They speedily
withdrew to a distance, for they knew not how many
would follow after, nor how dangerous they were.

"Let's call uncle," said Mary; "I am sure he
never saw such a nest as that."

"I'll run and tell him and Cousin Charlie, and
all of them," said Bertha. "Do you stay close by
and see that the bee doesn't go away."

Bertha fulfilled her self-appointed mission with
great promptness, and in a very short time returned,
accompanied by Uncle Samuel and her cousins.
Uncle Samuel, imagining what kind of a nest his
little nieces had found, took his portfolio along with
him, that he might be prepared to illustrate the
home of the insect without the trouble of making
drawings as on the former occasion.

Mary had ventured again very close to the little
mound, and as her uncle approached she pointed to
it and said,

"Just look, uncle! Did you *ever* see a nest like that? I am sure the bee must have just ever so nice and ever so many little chambers in its house for the baby-bees to grow in."

" Yes, indeed I have, Mary; and I am so glad I can tell you all about the history of this bee, and show you a picture of the nest, or of one built by a bee just like it!"

A very pleasant group soon found their positions on the grass close by the mound which was now the centre of interest. A more earnest teacher and more interested and attentive pupils could nowhere be found. Uncle Samuel was resting upon his right elbow, while his nephews and nieces formed a semicircle before him :

" The builder of this little mound belongs to a group of bees called *mining-bees.* Their scientific family name is *Andrenidæ.* They are ordinarily no larger than a house-fly. Some species of this group make their nest in very hard ground. This bee belongs to the species called *Andrena vicina,* and chooses grassy fields for the place of its nest.

" When the mother-bee wants to prepare a home for her offspring she makes a circular and deep hole in the ground a little wider than the diameter of her body, so that she can move down and up at pleasure. This burrow is sunk perpendicularly, with short passages leading to the cells, which are slightly inclined downward and outward from the main passage. Here is a cut (Fig. 43) which will convey the idea to your minds better than any description. The whole labor of digging the nest and providing food for the young is performed by the mother-bee. The males of the solitary bees are idle, and the females are unprovided with laborers, such as the moss humble-bee or the hive-bee has.

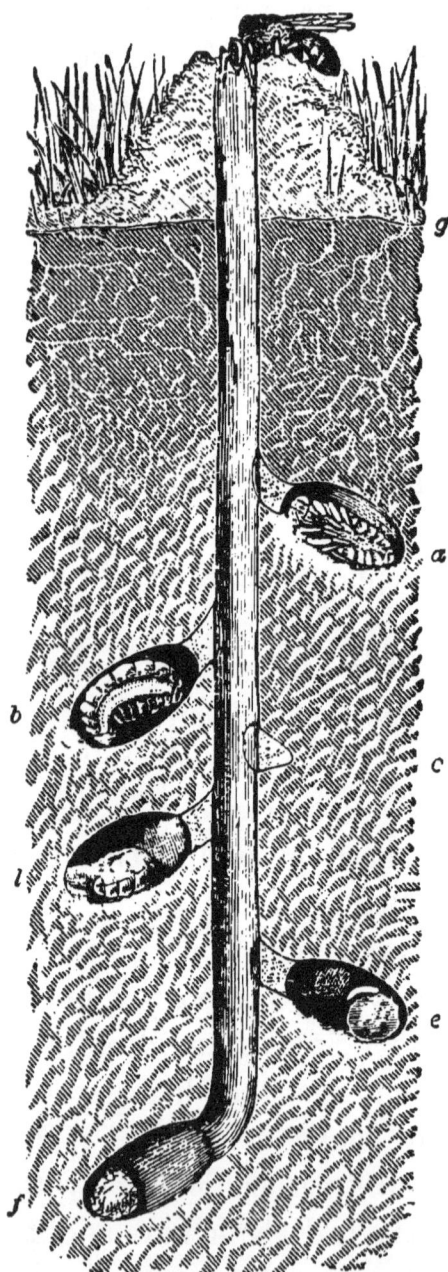

Fig. 43—NEST OF THE MINING-
BEE.

"The walls of the main passage—which might be called the long hall—are rough. The bee does not believe in outside show, as a great many people do, but she is careful to fit up very nicely the little chambers she has prepared for her young. They are made with smooth walls, and are lined with a varnish that, on hardening, looks like the glazing on earthenware. The depth of the long hall is usually from four to eight inches. Some species make their burrows ten inches deep."

" That is a very deep burrow for so small an insect," said Charlie. " If the bee is only one-third of an inch long, then it digs a great tunnel perpendicularly thirty times as long as itself. If a man six feet high would dig such a tunnel, thirty times deeper than he is tall, it would make a well one hundred and eighty feet deep."

" Your calculation is correct, and it proves that insects have greater physical strength and greater skill for execution than man, for man could hardly accomplish the digging of a tunnel so many times longer than himself without aid, while the mother-bee does it all herself.

"As soon as a cell is made ready for its inhabitant, and is supplied with pollen for the food of the larva, the mother deposits her egg and closes it up. She then proceeds with her work until she has made from five to eight cells. The most recent cell is, therefore, the one deepest down. In this picture (Fig. 43) the cell *a* contains a pupa; *b*, *l* and *e* contain larvæ in different stages of growth; *f* contains freshly-deposited pollen; and *c* is the beginning of a cell. At *g* is represented the level of the ground.

" The earth which makes the mound was scooped out by the bee while digging its gallery and cells. When the grub is hatched in its cell this earth is sometimes used to close up the passage to prevent the entrance of ants or other enemies."

"When going to school a few days ago," said Henry, "I saw a little opening in the ground like this one close to where I was walking. As there were no heaps of earth near the holes, I presume the bee had carried it all away, piece by piece, when she dug out the tunnel. I have seen the bee, and it does not belong to the same species with this bee, for it does not look like it."

"I remember the spot to which you refer. That bee belongs to the class of *upholsterer-bees,* and is another of the bee-mechanics, of which there are many. The common name of that bee is the *tapestry-bee,* or the *poppy-bee,* so called from its selecting the scarlet petals of the poppy as tapestry for its cells. But it has also a scientific name; for all the busy inhabitants of the insect-world, when they appear in the presence of the royal princes of science, who profess to be very learned, must be clothed in an appropriate court-dress, otherwise they are not recognized; so when our plain, every-day poppy-bee appears in the reception-room of the entomologist, it is introduced as *Anthocopa papaveris.* The first of these words means *flower-cutters,* and the second means *poppy.*

"The mother-bee digs a burrow in the ground about three inches deep, making it wider at the bottom than at the top, so that it resembles somewhat the shape of a common flask with a long neck.

The interior of this cell she polishes sufficiently to receive the gorgeous tapestry with which its walls are to be hung. It is no mean, unadorned chamber in which her offspring shall spend its early days. The mother has not only an eye to the comfort, but also to the beauty, of her nursery, and while its larder is supplied with the richest food, its walls are ornamented with tapestry of Nature's finest weaving.

"The walls being made ready, the skillful little worker goes forth in search of the scarlet poppy, and, resting upon the opening flower-leaves, cuts off a small piece of an oval shape, seizes it between her legs and conveys it to her nest. Entering, she descends to the bottom and neatly and tastefully spreads her carpet. This operation she repeats until the bottom is overlaid with three or four leaves. Having thus made the floor of her chamber soft and comfortable for its intended occupant, she hangs her tapestry on the walls, extending it even beyond the opening. If she happens to bring to her nest a piece of poppy-leaf too large for the place for which it was designed, she cuts off what is not needed and carries off the scraps."

"She must have very sharp mandibles," said Charlie, "when she can cut the petals of a poppy-flower, for I tried to cut some rose-petals yesterday with mother's sharp scissors, and I couldn't do it."

"Her scissors are always very sharp and she is very skillful in using them, so that she cuts them without a wrinkle, and spreads them on the walls of her house as smoothly as the most perfect paper-hanger.

"The little chamber being ready for occupancy, the kind mother goes forth again among the flowers to gather pollen and honey, which she mixes in proper proportions, and deposits within it as food for her offspring. When she has filled the chamber to the height of half an inch she lays an egg, and over it folds down the tapestry of poppy-petals from above. I have here a drawing which will introduce you to the interior of the poppy-bee's nest. (Fig. 44.)

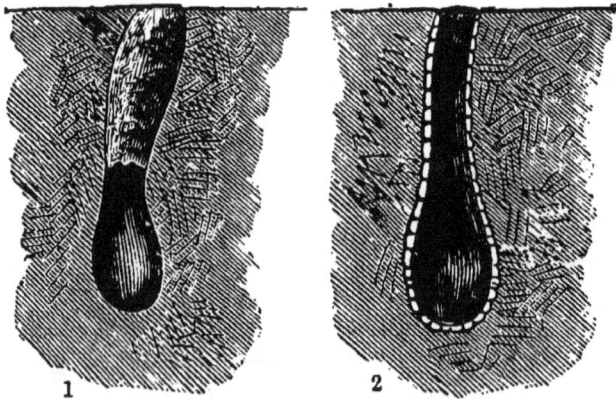

Fig. 44.
1. The cell before the egg is deposited within; 2. The red poppy lining depressed.

"In this cell the little tapestry-bee spends its early days, and undergoes those wonderful changes of form by which it becomes winged, and fitted by its new tastes and new organs to dwell in a world of light and enjoy a life of greater and more exalted freedom; and when it leaves its pupa-case, having no longer any love for its narrow prison with its walls of faded tapestry, it is moved by higher aspirations to find its way out of its enclosure and upward to the new life and brighter glories beyond.

"I must pause just here to say that this larva, feeding upon its pollen food in its tapestried chamber, is like man. The world is beautifully hung with the richest tapestry, and it is abundantly supplied with food suited to his earthly taste; but like the narrow home of the larva, it is dark and prison-like, and is not his abiding-place. Man, like the crawling larva, has a higher nature that cannot live on earth's richest provision—a nature that longs for a higher and better life. So when his food is all consumed and his earth-life draws to a close, the tapestry of his earth-home will fade, and his nobler nature will rise to bask in the sunlight of a brighter and happier world, to feast on daintier food and to move on untiring wing and for ever amid the beauties and glories of heaven's fields of light and life.

"But sometimes the ichneumon-fly lays its eggs in the nest of the tapestry-bee, and then its offspring never rises to enjoy the raptures of a new life; neither will you ever enter that better world, nor enjoy the higher life in glory, if you permit your greatest enemy, sin, to consume the life of your soul."

At this stage of the conversation Bertha's attention was attracted by a picture that had fallen out

of her uncle's portfolio, and was borne by the wind
to where she was sitting.   Picking it up, she said,

"Look, uncle, what picture is this?   It fell out
of your portfolio.   Here are two bees eating leaves,
and another flying to the branch."   (Fig. 45.)

Fig. 45.—Rose Megachile (*Megachile centuncularis*)

"That represents another one of the solitary bees,
belonging to the leaf-cutters.   Its family name is
Megachile, and its specific name is *centuncularis*.
Its name is much larger than itself.   Its habits are
very curious.   It is said that a French gardener,

finding some of the nests of this insect in his garden, and believing that some witch had made them, sent them to his master to know what he should do to keep away the witches. His master showed them to some learned men, and soon discovered that an insect had built them, and so the fears of the gardener were dispelled.

"Let us imagine ourselves looking in upon the mother-bee while she is engaged in preparing and constructing her nest. She has chosen the limb of a decaying willow tree, and bores a hole in it horizontally. She prefers this tree when she can get it, and for this reason she is sometimes called the *willow-bee.* When she can do no better she makes her nest in a beaten pathway, and sometimes in the cavities of walls. When our bee has bored a hole in the wood for seven or ten inches, she goes forth to collect materials for the structure of her cells, and generally chooses the leaf of the rose-bush. On this account she is often called the *rose-leaf cutter,* or *rose-cutter bee.**

"The process of cutting out the pieces of the leaf that are to form the walls of her cells is very curious. How expeditious she is! She has no time to lose. Hovering over the rose-bush for an instant, as if reconnoitring the grounds, she alights upon a leaf, taking her station upon the edge, so that the margin passes between her legs. With her mandibles, keen and sharp, she cuts out a circular piece

* This is a European species.

with more neatness and despatch than any one of you could do with a pair of new scissors. Just before she cuts the last fibre she balances herself on her wings and bears off in triumph the separated piece as soon as it parts from the leaf. Holding it in a bent position, perpendicularly to her body, she enters her nest and ingeniously applies it to the interior without paste or glue, knowing that the elasticity of the leaf will hold it to its place.

"The interior surface of each cell consists of several layers of leaf made narrow at one end, but gradually widening toward the other, till the width equals half the length. The several pieces of leaf are so laid that the serrated, or sawlike, margin of each is on the outside, as you see in the picture.

"It requires from three to five layers of leaf to make up the walls of each cell. In making each coating the bee is careful to lay the middle of each leaf over the margins of those already laid, so as to increase the strength of the wall. She is also careful to give the closed end of the cell a convex shape.

"When she has completed a cell she flies away to some cluster of thistles, that she may gather pollen from their flowers out of which to manufacture a rose-colored paste as food for her offspring. This she deposits in the cell in sufficient quantity, and then lays her egg, and closes all by fitting three pieces of leaf exactly circular, and which are cut out by the insect with unfailing accuracy, at a distance from the cell, and of the exact dimensions.

"In this manner she proceeds to construct cell after cell, the lower end of the one fitting into the open end of the next below it. This industrious little insect is not easily discouraged, for if by any accident her labor is interrupted or her edifice deranged, she loses no time and spares no pains in putting all things to rights again.

"Before changing into the pupa state the worm or larva spins a slight silken cocoon about the walls of its dark chamber, and thus fits it up for the scene of its last and most interesting change."

During all this conversation Charlie was very attentive and thoughtful. It seemed now to be his time to speak, and, straightening himself up, he said,

"Your history of these very interesting insects has made me feel ashamed of myself. I think of how many times I have given up trying to study out a difficult passage in Latin or Greek just because I could not understand its construction at once. I lack patient perseverance in toiling. The little mining-bee carries pellet by pellet up the sides of its burrow till it has dug for its young a nest thirty times as long as itself. The other bees you have told us of only accomplish their work by perseverance. Only a little at a time can any of them do, but they are always busy, persevering and determined. Ought we not to learn from these industrious insects to be persevering in doing whatever we have to do? Their motto seems to be,

'Never give up!' That shall be my motto too.
I will not let a little bee be more noble and persevering than I am. The willow-bee, that despairs
not though accidents destroy her work, calls to my
mind that little poem of Tupper I learned out of
my school Reader, the last verse of which is—

> '"Never give up!" if adversity presses:
> Providence wisely has mingled the cup;
> And the best counsel in all our distresses
> Is the stout watchword of "Never give up!"''

"There is another lesson," said Uncle Samuel,
"that you must not fail to learn from these facts,
and that is the skill and ingenuity of these little
operatives. Surely the mechanics among insects
are not to be despised because they are diminutive
creatures. The poppy-bee and the rose-leaf cutter
carry no plumb-line, compasses or square, and yet
with unerring accuracy they cut the materials with
which they construct their dwellings, without making a single mistake, just as the stones of Solomon's
temple were fitted to their respective places before
they were put upon the wall. Is there not in all
this a power manifested that excels human reason
and art? This wise calculating forethought cannot
be the offspring of blind, undesigning chance. It
is the Almighty's teaching. He it is that has so
admirably proportioned the insect's knowledge and
skill to her necessities. 'Let everything that hath
breath praise the Lord. Praise ye the Lord.'"

CHAPTER XII.

THE next day after the conversation which made the subject of the last chapter Henry and James were in the loft of the carriage-house, when James observed near to the lower end of one of the rafters some patches of mud of a light-yellow color. They had learned to suspect insects of being the architects of every strange little structure with which they met. So no sooner had James made the discovery than he called to Henry :

"Harry, come, I have found a nest of some insect. Look at these little patches of mud. See how carefully they are built up. I'll warrant some insect that has wings has put them there, for how else could mud get up here?"

"Give me your knife, James, and I will try to

165

loosen them from the rafter, and we will take them to uncle. He will tell us all about them," said Henry.

By care they separated one, and agreed, although it was a small one, not to try any more until Uncle Samuel should see it and tell them how it got there. They soon found their way to his study, and laid their prize on his table.

"Ah!" said Uncle Samuel, "you have been robbing some insect of its home. What cruel boys you are! How would you like it if some strong being, ever so many times larger and stronger than man, should carry off this house and all that are in it away to some very large room, just to have it pulled to pieces to find out what it contained? Wouldn't you think it very cruel?"

"I think it would be *very* cruel to treat *us* so. But then insects are not our neighbors, and we are not required to love them as we do ourselves," said James.

"Uncle," asked Henry, "have you forgotten a verse in the Psalm you read this morning? I remember it: it is, ' The works of the Lord are great, sought out of all them that have pleasure therein.' Now, this little house is part of God's works, and we have pleasure in seeking out the wisdom and skill he has given to the little insect that has built it. That is the reason why we have brought it to you, because we think you will tell us all about it. Then, perhaps, there is no insect in it. It may be an abandoned house. But you know we can't argue with you, and you are just trying us, that's all."

By this time the other children, seeing James and Henry running to their uncle's room, had gathered in, for they thought that something new must have been found, and they did not wish to lose any of their uncle's stories about insects.

" You reason very well," said Uncle Samuel, " for *young* philosophers, and perhaps I shall not succeed in making you repent of wrong-doing in carrying off this insect-home; so I will proceed to give you some account of the little architect whose skill is displayed in this rough-looking structure. But first let us open the cells which this home contains. There, I have cut into one and find a worm three-fifths of an inch long, surrounded with a dark-brown silken covering closed perfectly on all sides. It seems to be awaiting some future destiny, and is quietly and calmly resting in its solitary home till its change comes. But we have destroyed its habitation and dislodged it from its bed, and its future prospects, whatever they may have been, must be unrealized. Now let us open another."

" Oh, just look !" said the astonished Mary. " What a nest of spiders ! And pretty ones too ! Is it a spider's nest ? Let me count them. One, two, three, fifteen ! The dear little things ! Isn't it a pity to disturb them !"

" I see a little white worm," said Bertha—" oh !

ever so little! I should think it would be afraid to
be shut up there all by itself, with so many spiders
bigger than it is; I should. I do wonder how it
got there, and how the spiders shut themselves in?
If I were a spider I wouldn't like to be shut up
and packed into such a little dark room. I wonder
what so many of them do there, anyhow?"

The contents of the little cell were a mystery to
all the children. Even Charlie acknowledged himself puzzled unless the little worm fed on the spiders; but this he thought could hardly be, because
it was too small to eat so many, being only about
one-sixteenth of an inch long. When they had
exhausted all possible conjectures without satisfy-
ing themselves, Uncle Samuel said,

"This is the work of the dauber-wasp—a solitary wasp whose instinct in constructing its habitation and providing for its offspring has always
excited the admiration of the naturalist. It belongs, like the bee, to the *Hymenoptera*, and its
scientific name is *Pelopœus lunatus.* The English
of the generic name is '*mud-maker*,' and the specific
name means '*mooned.*'

"Here is a picture of a wasp that sometimes
lays its eggs in the dauber's nest. It is called the
black wasp (*Trypoxylon albitarse*). (Fig. 46.) I have
here also sketches of 'mud cells' drawn from nature
and made by four distinct species of a genus or kind
of mud-daubers called *Agenia*, which means 'without (illustrious) birth.' Alongside of the mud cell

(*d*) there is a picture of the wasp that constructs it. It is called *Agenia bombycina.* (Fig. 48, *b.*)

"I have here a picture of a grand wasp, which I must show you. (Fig. 49, p. 171.) It is called by Say *Stizus grandis*, or the gigantic digger-wasp. It is a native of Pennsyl-

Fig. 46.—Trypoxylon albitarse.

vania, and is called the digger-wasp because it burrows in the ground, and in the burrow lays its egg, and then deposits locusts in it for the larva to feed upon.

"Here are sketches, drawn from Nature, of the 'mud-cells' built by three distinct species of the

*Fig. 47.—*Mud-celled Nests of the *Agenia* Wasps.

Agenia. They represent the natural size of these remarkable nests, and, as you see, their parts are connected together like a string of flattened beads.

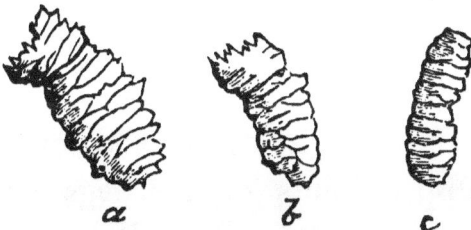

The smallest cell (*c*) is built by a very small wasp (*Agenia subcorticalis*), so called because it constructs its little home always under the loose bark of standing trees. The middle cell (*b*) is the work of a larger wasp (*Agenia architectus*), and it is found

sometimes under logs and stones, and sometimes under the loose bark of trees.   The cell (*a*) is constructed by the *Agenia mellipes,* and is found exclusively under the bark of standing trees.

" All these wasps build and provision their nests alike, so that what I tell you of the instinct and habits of the common dauber-wasp will do almost equally well for the *Agenia.*   The form of the *Pelopœus lunatus* is peculiar, as it has a very small abdomen, separated from its thorax by a

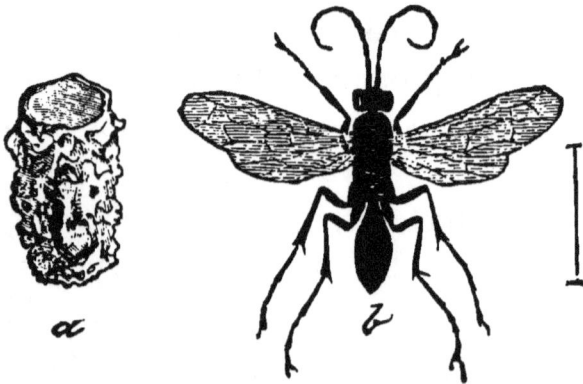

Fig. 48.—*Agenia bombycina.*   (*a*) Its nest.

long foot-stalk, sometimes bright yellow and sometimes black.   The upper joints of the legs are black and the lower joints are bright yellow.   It has two antennæ, which it rolls up spirally; their color near the head is yellow, but black throughout the rest of their extent.   Its wings are a dark brown, terminating in a bright maroon.   Its thorax (or chest) and abdomen are of a deep brown-black, softened by thick downy hair on the thorax.

" The mother-wasp, when ready to build her nest, seeks some sheltered spot in which it may be protected from the rain ; hence her cells are usually formed upon rafters or in the upper corners of windows in rooms but little used. She then goes forth in search of clay suited to her purpose. This she moistens with cement of her own manufacture, adding particle to particle until her load is oftentimes larger than her head. Taking it up with her jaws, she flies away to the chosen spot and deposits it in its proper place, laying the foundation of her

Fig. 49.—THE GIGANTIC DIGGER-WASP (*Stizus grandis*, Say).

future home. Load after load she carries, prepared in the same way, and shapes it so as to adapt it to the place she intends it to occupy. The cement keeps the clay soft and pliable, so that she can mould it as she pleases, and when it dries the clay is almost as hard as a stone. In building the cell she is careful to preserve its cylindrical form and to make the sides even and smooth.

"When she has completed the first cell the mother-wasp goes forth to seek food for her offspring. It is not among the flowers she goes, as the bees did about which we talked yesterday. The little baby-wasps would starve if born among the pellets of sweetened pollen. She goes among the spiders, and, carrying them off one after another, she places them in the cell until it is filled. As soon as she captures a spider she stings it so as to render it helpless, but not to kill it. She then takes one of her fore legs in her mouth, and, embracing the body with her fore and middle legs, she carries it off in triumph. When the cell is filled she closes it with mud.

"Another cell is built in the same way, parallel with the first, then another and another, until she has made all the cells she has designed to make. These she covers over with clay or sand, so as to make them still more secure against the attacks of insect-enemies that seek the life of the helpless inmates of the cells.

"As the cells are not all made and occupied at the same time, the grub which inhabits the oldest cell will come to maturity first; so that in a nest made up of several cells the insects will present at any one time different stages of growth. I have here a picture taken from a photograph of a nest which I took from a rafter in an unoccupied garret.

In removing this nest the lower cell was broken off, so that only a portion of its walls is represented in the cut. (Fig. 50, *g.*) The upper cell, *a*, is the oldest, and when the photograph was taken, the thorax and abdomen were black, and the eyes were perfect. The middle wasp, *b*, was yellow in all its parts, and the eyes were not fully developed. The lower one, *c*, had just changed from its worm shape, so as to exhibit

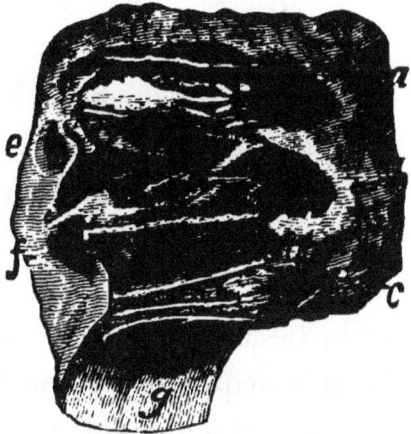

*Fig. 50.—*Wasp-Grubs in Cells.

the long peduncle between the thorax and abdomen; the legs and eyes were wanting. There are two openings (*e* and *f*) shown on the side of the picture, which belong to other cells lying back of those represented.

"When this nest was removed, I took away a portion of the cells and the silken cocoon which enveloped the insect, so that I might observe any change that might take place. The lower wasp was then a worm without eyes, and terminating in a sharp point at each end. For some time I examined them daily, but noticed no change in the worm. At length the time of transformation came. And how remarkable it was! The grub which I had seen but a few hours ago legless, headless, eyeless, now presented to me abdomen, peduncle, thorax, head and the beginning of eyes and legs. How changed! Can it be, thought I, that this is the *worm* I so lately looked upon?"

"I suppose," said Charlie, "God did not give the little worm any eyes or legs, because it couldn't use them in its dark narrow cell, where all its food was provided and it had nothing to do but to eat."

"You are right in your conjecture, Charlie. In each stage of the wasp's life it is abundantly provided for; it lacks nothing when a grub, and it would be as useless to give it eyes then as wings. When it acquires all the parts it needs to fit it for motion in the air, and the duties and pleasures of its higher life, it does not remain any longer in the home of its childhood, but eats its way out, and sees and flies and walks just as if it had always done so.

"There is another very interesting fact I must tell you. When the mother-wasp is building her nest in a room, she always takes the most direct course leading to it from the clay-bed which provides her with material. But if the passage into the house first chosen is closed, she flies round the house until she finds an opening, and entering there directs her way through other rooms, without erring, to her nest. Is there not something like intelligence in such an act as that? She evidently possesses a faculty by which she can adapt herself to circumstances, and, as though gifted with reason, when she is shut out from reaching her nest by one passage, she seeks to find another passage-way by which she can reach it.

" Is not insect-life becoming more and more wonderful as you progress in its study ?   How marvelous is the providence of God, who does not neglect the wants of the smallest of his creatures!   His care extends to the little eyeless grub that is born and grows in such mud-cells as these, providing for it until it is elegantly equipped with the most beautiful wings, and eyes of wonderful structure, and an instinct that rivals the power of reason and astonishes us with its exploits.   You must never despise the rough, inelegant infant-home of the dauber-wasp.   It speaks to you of God as clearly as does the starry sky.   It tells you to be content to 'make haste slowly,' doing well and thoroughly everything you undertake."

CHAPTER XIII.

AS the heat of summer grew more intense, the ants began to make their appearance, much to the annoyance of the household, and our young naturalists were very anxious to know something of their history. So it happened that when walking in the woods with their uncle on one of their rambles, James stumbled on an ant-hill, and produced no small disturbance within the industrious community. This was all that was needed to call forth from Uncle Samuel a brief account of their life.

"I have often wanted to ask you to tell us the history of the ants," said James; "and now that we have disturbed a nest of them, we will all sit down in this beautiful shade and listen to you."

"That we will," said Mary, who was not a little

wearied by her long walk, and hence quite ready for a rest.

"Oh, do tell us something about them," said Bertha, "for we have *so* many at our house in the city. They just get into everything, and mother says they are the greatest pests in creation. They are so busy too—always at work as if they had more to do than anybody else. Aunt can't keep them out of her sugar, and they get into her milk and *everything*. I wonder what they are made for, anyhow?"

"I can tell you, cousin, one thing they are good for," said Charlie. "They teach us lessons of industry and perseverance. Just look at these ants whose nest James stumbled on. How busy they are! They have already commenced to repair the damage he has done their home. Don't you remember the lines that Dr. Watts wrote about the ants?—

'They don't wear out their time in sleeping or play,
But gather up corn in a sunshiny day,
And for winter they lay up their stores;
They manage their work in such regular forms,
One would think they foresaw all the frosts and the storms,
And so brought their food within doors.'

But uncle knows all about them, and I will cheerfully resign the office of teacher to him and become a listener."

By this time they had all comfortably seated themselves on the grassy lawn, shaded by oaks and

12

maples, and Uncle Samuel, laying his portfolio by his side, began :

"You will first want to know to what order of insects the ants belong. Seeing that the busy inhabitants of this ant-hill are wingless, you will not be able to classify them. But ants have wings during a part of their life, and these wings are membranous."

"Then they belong to the *Hymenoptera?*" said Henry.

"You are right," answered Uncle Samuel; "and as the Latin word for ant is *formica,* the ant family is called *Formicidæ.* It is a very large family, and is made up of a great number of species. Each of these species has some peculiarity by which it is distinguished from all the rest, while there are a great many traits of character which are common to the whole family. I will first tell you about some of these common characteristics, and then I will give you the history of a few individual species.

"Ants are among the most wonderful of insects. The instinct of the hive-bee is certainly wonderful. You can hardly look at a honeycomb filled with pure, clear honey without admiring the skill with which the cells are made and the process by which such delicious sweetness has been taken out of the cups of flowers and placed in the cells. The instinct, almost rivaling human reason, which the insects about which we have already talked so frequently display, has astonished you, but our little ant surpasses them all in industry and wisdom. The great king Solomon, disgusted with the idleness and worthlessness of the idler, said to him, 'Go to the ant, thou sluggard; consider her ways and be wise: which having no guide, overseer or ruler, provideth her meat in the summer, and gathereth her food in the harvest.' This same king gives to the ants the first rank among the four things on the earth that are wise. It might be that this 'king, who was the wisest king in the world when he reigned, derived a great many suggestions of wisdom from the study of the manners, customs and laws of the ants."

"Why, they are not people, uncle," said Bertha, somewhat astonished at his last remark. "You speak of them just as our geography speaks of the different nations in the world. Do you mean that the ants have a government as men have?"

"Yes," answered Uncle Samuel, "they live in communities, and as they all work for the common interests, they govern themselves by certain laws which are known and obeyed by all. Ants are positive republicans. They do not believe in monarchy, as the hive-bee does, unless in the case of a nation of ants that live in Africa that have a monster queen, for whose interests the whole community lives. In their wonderful republics mobs are never known. All are obedient to authority, and each one has special duties to perform, which he discharges with great pleasure and promptness. They are very patriotic, and if their habitations are attacked, they defend themselves with great heroism, and generally with success.

"That all the departments of their model republics may be well sustained they are divided into three classes—males, females, and workers or neuters. Of these the males and females have wings, while the workers are wingless. Each one of these orders is confined to its respective duties in the community as definitely as the castes are in India. They never contend about woman's rights. All parties are satisfied with the work and the position assigned them by the laws of their community, and live together in the most perfect harmony and affection.

"Ants vary greatly in size, some being very small, as the common red ant which infests our cupboards and sugar-closets, while others are an inch long. Their head is broad and their hind-body is large. The general form of the body is slim. They have six long legs. They have antennæ, but in the form of an elbow often. With these they examine everything they meet, and hold intercourse with each other. Their jaws are very strong, and they can use them for pincers, tweezers, scissors, pick-axe, fork or sword, according to circumstances. Each pair of legs is armed with a spur and fringed with very short hairs, which are their brushes. The eyes of the males are large and prominent, whilst the eyes of the females and the workers are small. The workers are smaller and stronger than the males, and these are smaller than the females. The males have four wings. They die after the pairing season. The females cast their wings about the time that the males die. The females lay their eggs in parcels of half a dozen or more. These are taken by the workers and deposited in a safe part of the nest till the young grub is hatched. The grub is white, footless, with a horny, brown head. It is the office of the worker to supply the grubs with food and to watch them carefully until they are full grown. At adult age they spin for themselves white silken cocoons and pass into the pupa state. These are what the common people in Europe call ant-eggs, and they are collected and sold in some of their cities as excellent food for mocking-birds and nightingales."

"Why have not the neuters, or workers, wings as well as the males and females?" asked the inquisitive Mary. "Are they to do all the drudgery, and then have to go everywhere on foot? Poor little worker! I pity it."

"If the workers *needed* wings, I have no doubt God would have given wings to them. But what would carpenters and masons do with wings? And such are the workers. They are the carpenters and masons that build their houses, the ant-hills. They are the grocers who see to getting the marketing needed by the whole community. They are the nurses who take care of the young and rear them up to anthood. That they may not wander from home or neglect the important interest committed to them, they are without wings; hence they become very much attached to the community and faithfully attend to their duties. Their intimate relation to the members of the ant-nation makes them their natural guardians, while their authority and their strength enable them to defend the republic in the time of war."

"Do ants ever make war with one another?" asked James.

"I think they do," said Charlie. "I read, a few days ago, in a book written by a German author, a Mr. Hanhart, an account of a battle witnessed by him between two hills of brown ants and the inhabitants of three hills of black ants. The black ants were smaller than the brown ants, but much more numerous. The brown ants made the attack, moving in a line of battle twenty-four feet long. The black ants marched out to meet them. When they met the battle commenced. Their weapons were their jaws, stings and store of poison. After the fighting commenced the parts of the bodies, feet, legs, antennæ, abdomens and heads were strewed all over the battle-field. At length the brown ants were conquered and fled, leaving their homes and fortresses in the hands of the enemy."

"Such stories are told by all careful observers of the customs of the ants," said Uncle Samuel. "It would be well for the human race if the same cruel and barbarous stories could not be told of civilized nations. Wars should be confined to insect tribes and other orders of animals, while man's superior energies should be devoted to the culture and development of his higher and God-like nature."

Mary, who always became restive when the conversation became too deep for her understanding, diverted her uncle from his moral lecture by the questions,

"How are ant-hills made? and what are they used for?"

"Ant-hills are built, sometimes, with the dirt which the workers take out of the tunnels and chambers of the nest. Sometimes the hills are composed of fragments of wood, pieces of straw, dry leaves and remains of insects. Each species has a plan of house-building peculiar to itself. The hills are the homes of the ants. Here they eat, sleep and have their domestic joys. Here they rear their offspring. The hillock is nothing more than the covering of a many-chambered palace under ground, all fitted up for the convenience of the ant-family which occupies it. The entrance to the nest is simply a hole or tunnel which the thrown-out rubbish conceals. As the excavation goes on, the tunnel branches out into a labyrinth in all directions, and the pile of earth grows larger. The interior consists of corridors, landings, chambers and spacious rooms communicating with each other by long and short passage-ways. All of these passages lead to one central and grand gallery higher than the others—a kind of council-chamber, where the ants hold all their mass-meetings.

"I presume that you have some idea of the construction of the homes of the ants already, and your imagination can picture its passages and its rooms; but here is an engraving that may give you a new thought about our busy little friends. It is a magnified picture of an ant drawing milk from one of

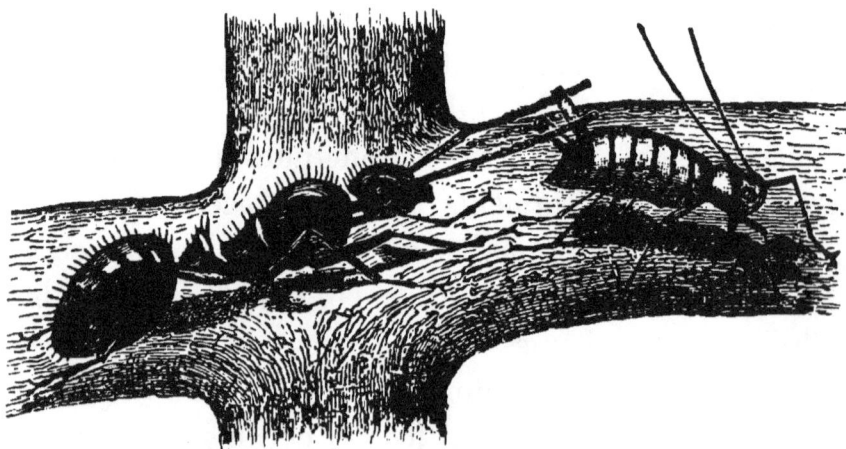

Fig. 51.—ANT AND APHIS.

the little plant-lice (*aphides*) that so often annoy our rose-bushes."

"Milk from plant-lice! Now, uncle," said Bertha, "you are just making believe. Plant-lice are not cows."

"I am not surprised that you doubt what I say, for the fact is very strange indeed; but you must be willing to believe very remarkable things when you are studying the history of insects, for it is full of them. Plant-lice draw from the vegetables on which they feed a juice which they convert into a peculiar kind of liquor called honey-dew. The ants know this fact, and when they discover plant-lice on any vegetable or tree the whole community turn out, and they persuade the little insects, by gentle touches of their antennæ, to disgorge the sweet juice drop by drop, which they immediately appropriate to their own use. The scientific name of plant-lice is *aphides*. They are homopterous insects, and have a remarkable history. They are very injurious to the plants on which they feed and they generate with wonderful rapidity, the mother of the first generation in one season becoming the progenitor of ten thousand millions of millions. A French writer calls them 'the milch-cows of the ants.' The ants prize them so highly that they are sometimes taken into their nests, where they are carefully watched and kindly treated for the sake of the honey-dew they secrete; so you see that ants hold property in stock as men do. Sometimes war arises between two ant-republics for the possession of these aphides. When such a war takes place the innocent aphides are made to suffer many wrongs, even from those who pretend to be their friends.

"But I must not forget to tell you something more about the domestic habits of the ants. The females live together in great harmony, and having transferred to the workers the entire care and supervision of their young families, they have quite

an easy time of it. The careful nurses tenderly
provide for the comfort of the charge committed to
them. During the day the grubs are placed in the
open air to get the benefit of the sunlight. That
the nurses may know when the sun shines on the
hill, sentinels are placed just under the roof of the
nest to observe the rising of the sun. When they
are assured of the fact, they hasten to those who are
watching by the cradles of the young, and, by touching them in a significant way with their antennæ,
inform them that the sunbeams have reached the
surface of the ant-hill. In a few seconds all the
avenues are crowded with workers carrying the
larvæ in thin cocoons out of the nest to put them
on the top of the hill, exposed to the sun's rays.
When the heat becomes too great, their tender
guardians remove them to chambers close to the top
of the hill, where the heat of the sun is more mild.
When the pupæ are ready to escape from their cocoons the workers tear open their silken covering,
and thus assist them in escaping. They then watch
over the newly-born, feed them, teach them to walk,
and never abandon them till they can help themselves.

"Some kinds of ants capture and hold slaves.
This fact is so remarkable that it would not be believed if it had not been discovered by most reliable observers, such as Francis Huber. I find in
my portfolio pictures of two species of slaveholding
ants which belong to this continent. The first group

represents the russet ants, in science known as *Poly-ergus rufescens.* (Fig. 52.)

Fig. 52.—RUSSET ANTS (*Polyergus rufescens*).

" The next group represents the blood-red ant, *Formica sanguinea.* (Fig. 53.)

Fig. 53.—*Formica sanguinea.*

" The slave-ants are the ashy-black (*Formica fusca*) and the mining-ants (*Formica cunicularia*). The former are more readily attacked, because they are more easily conquered. The latter defend themselves with more vigor and success. Here is a group representing the male, worker and female of the mining-ants. (Fig. 54.)

Male. Worker. Female.

Fig. 54.—MINING-ANT (*Formica cunicularia*).

" The russet ants have jaws made for war; they appear to be well trained in the military art, and fight with great ferocity. They are so dependent on their slaves that without their constant service they would all die in a single year. The males and females do no work, and the workers are trained to no occupation except that of capturing slaves. They can neither build their own nests nor provide food for their young. If at any time it is necessary to abandon their nests, the slaves actually carry their masters and mistresses in their jaws. How wonderful this instinct! What a remarkable instance of the influence of slavery in making lazy and helpless creatures of the masters!

" The blood-red ants are less ferocious and not as good warriors as the russet ants. They regard their slaves as household servants, and treat them with much more kindness than those just described. They have fewer slaves, and aid them in building their nests and in other household duties. If their dwelling is attacked by an enemy, the masters show their esteem for their servants by carrying them down into the lowest apartments as a place of the greatest security. The slavery that exists among the ants is attended with no cruelty toward the slave-ants, inasmuch as they perform the same services in the nests of their masters as they would have done as freemen in their native community.

" In Western Africa there is a remarkable species called the driver-ants. Their scientific name is *Anomma arceus.* These ants are exceedingly ferocious. They do not live in permanent houses as other ants do, but go everywhere in search of prey. They travel during the night or in cloudy weather, the direct rays of the sun being fatal to them. If compelled to travel during the day when the sun is very hot, they construct arches of clay, using cement which comes from their mouths. If on their march they meet with a river, they make a suspension-bridge out of their own bodies, over which the main army passes safely. I have a drawing in my portfolio which represents a column of driver-ants on the march. The ants have ascended a tree in their course standing on the verge of the stream, and are descending a living ladder hanging from a branch and extending across the water. The covered way is also seen in the foreground, and a few of the larger ants are drawn of the natural size. (Fig. 55.)

" There is an ant living in Texas called the agricultural ant by the common people, and *Myrmica malefaciens* by men of science, whose habits are very remarkable in their way. It is said of it that it builds paved cities, constructs roads and sustains a large military force. It is a large brownish ant, and is called *agricultural* because of the attention it gives to the culture of the grain on which it feeds. In dry ground its house is covered with a low circular mound rising in the centre to the height of three, and sometimes six, inches. If its location is on low or flat ground, it elevates the mound to fifteen or twenty inches. In doing this it is guided by an instinct that foresees the danger to which its

habitation might be exposed by the overflow of the
low land, and it provides for its safety.

Fig. 55.—DRIVER-ANTS. (From Wood's *Homes Without Hands.*)

" Around the mound the ant clears the ground,
and levels and smooths it to the distance of three
or four feet from the gate of the city.   Within this

area it sows the seed of the ant-rice. This it tends and cultivates carefully, and when ripe it gathers the seed, a small, flinty grain, chaff and all, into its cells, where it is freed from its chaff and stowed away. If it should become damp and sprout, it is exposed to the sun and dried. The good grains are then taken back for use, while the damaged grains are left to decay."

"You astonish us, uncle," said Charlie, "with your stories about ants. They are certainly a remarkable class of insects. Why, I begin to think that man is not the only animal in this world that has been endowed with reason. The driver-ant and the agricultural ant both seem to reason from cause to effect. If not, how could they so wisely make provision against threatening dangers or for overcoming unforeseen obstacles? How small man seems to be when brought in comparison with the little insignificant ant! What immense buildings it constructs compared with its size! What grand council-chambers, what immense nurseries, what large and deep tunnels, what covered ways, what bridges, does it construct! Its tools—how simple and how skillfully used! Your facts have thrown in the shade the builders of the great pyramids and the grandest cities on the globe. I think I can never become vain or proud because of any work that I shall ever do.

"I think we ought all to learn from the ant not only to be industrious, but to work according to a plan, and that each one should do the part that belongs to himself—that is, that everybody should mind his own business, and not meddle with other people's business."

At this stage of the conversation the supper-bell called the little party in from their ramble to a plentiful repast, which they all enjoyed the more because of the interesting feast served up by the side of the ant-hill.

CHAPTER XIV.

IMMEDIATELY after tea Uncle Samuel was attended by his nephews and nieces to his study.
They had become so attracted by his wonderful stories about the ants that they begged him to resume
them.

"I know you can tell us more about the ants,"
said Mary. "Our teacher told us that there were
ants in the West Indies that entered people's houses,
and that everybody had to run before them or be
eaten up. Is that true, uncle? I should like to
see an ant that I would run from. I would just
put my foot on it, and then what could it do?"

"Oh yes," answered Uncle Samuel, "you would
be very bold if you had *one* ant only to contend

with; but suppose they should come upon you by the million, and cover you up and crawl into your ears and eyes and nose and mouth, and bite and poison you all over, what could you then do? That is just what the visiting ants that your teacher referred to would do with you if you were in the West Indies, and would not fly from the house into which they entered. This ant is called by scientific men *Atta cephalotes.* Here is a sketch of it. It is a savage-looking fellow. (Fig. 56.)

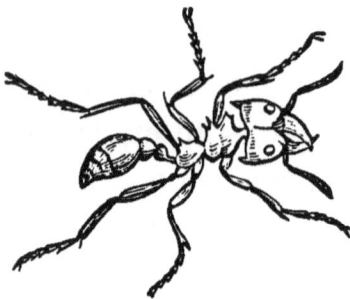

Fig. 56.—Atta cephalotes.

"These ants are as large as a common wasp, and are very destructive in their ravages. Once every year they leave their homes and go out on their visiting-tours. They are not select as to the rank of the persons into whose houses they enter as visitors. They treat all persons in their way alike. Their swarms are innumerable. They enter houses without knocking or any warning, run through all the rooms and kill all the large and small insects that they find—such as spiders, centipedes, scorpions, lizards and toads. If the human inhabitants do not leave the house to their exclusive possession, they will be devoured as well; so they are glad to flee before them. These ants are regarded as friendly to man, because they free all dwellings from the numerous pests which are so annoying to the inmates. So when the visiting ants have left a house the occupants return to find it in a much better condition as a comfortable home than before their arrival. They also remove noxious carrion, which would soon breed pestilence, and thus they contribute to the health of the entire community. Their superabundance, which would soon overrun everything, is checked by their enemies the ant-eaters and armadillos. These animals feed altogether on ants, and fix a limit to the increase of these destructive insects."

"Ain't I glad I don't live in the West Indies!" said Bertha. "How I would run if I saw them coming to our house!—So would you, Cousin Mary. They are not like our ants."

"Are not the white ants in Africa, that build huge hillocks, more to be feared than the visiting ants?" asked Henry. "I read something about them in my Reader in school."

"I am glad you mentioned the white ants, for I want to tell you that they are misnamed *ants*. They do not belong to the same order of insects. They introduce us to an order unknown to you. If their wings are examined, they will be found to differ in their structure from those of the true ant. The wings of the latter are membranous, but are not covered with nervures; while the former have wings which are membranous, but so thickly set with veins as to look like network. As these veins are called nerves, the insects that have such wings are called *Neuroptera*—a word derived from two

Greek words,* meaning *nerve-winged*. To this order belong the white ants—TERMITIDÆ; the dragon-
flies—LIBELLULIDÆ: the may-flies—EPHEMER-
IDÆ; and the ant-lions—MYRMELIONIDÆ. The
first three do not undergo complete metamorphosis,
as the butterflies and true ants do. There is not
much difference between their larvæ and the perfect insects, and they are all active during their
pupa state.

"I have in my collection of drawings represen-
tations of a species of white ants that at one time
occasioned considerable alarm at Rochelle, France,
because of their ravages. These drawings show the
inhabitants of a white-ant hill, and will help you
to understand more clearly their manner of life.
(Figs. 57, 58.)

"They unite in societies composed each of a great
number of individuals, and they live on trees and
in the ground, often attacking the woodwork of
houses. When they attack woodwork they make
numerous tunnels, and are very careful not to pierce
or deface the surface of the wood. Hence the furniture in houses, thus destroyed, seems perfectly sound
when by a very slight touch it will fall to pieces.

"You observe a great difference in the form of
the termites. The winged ones are the males and
females, the kings and queens of the ant-hill. These
crawl to the door of their house and fly away, coming to the ground after a short time. When a male

* *Neuron*, a nerve; *pteron*, a wing.

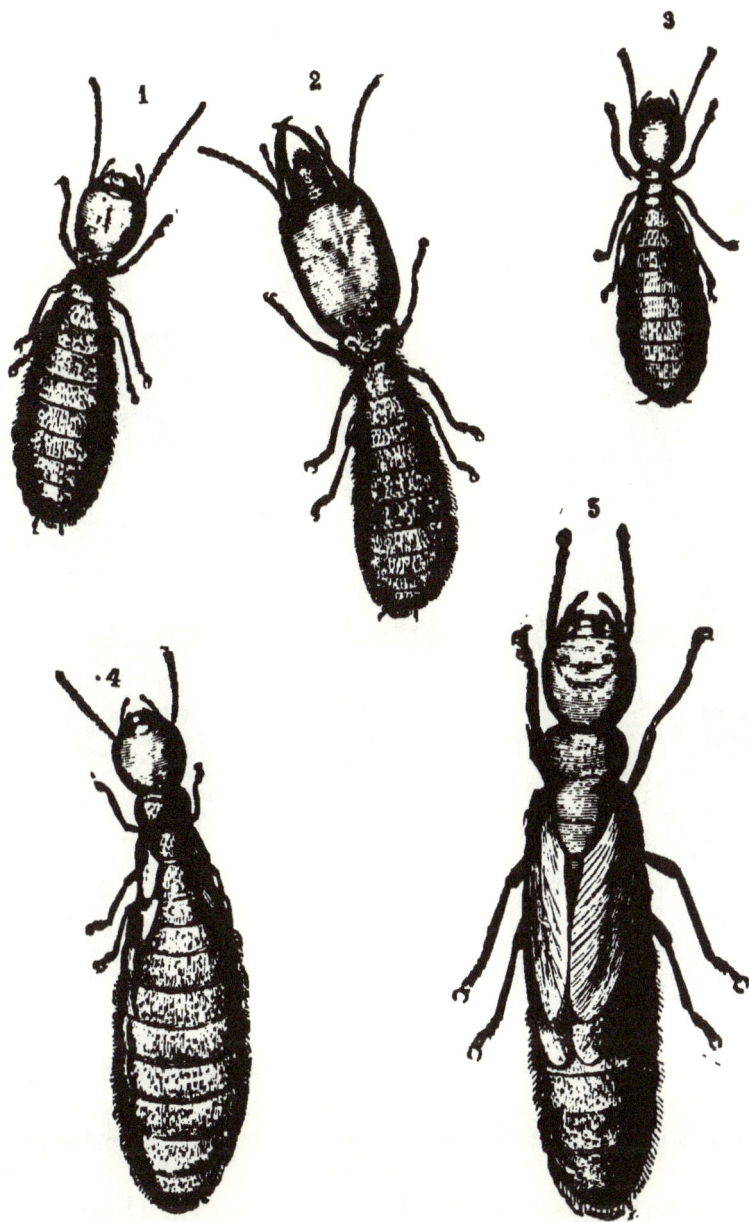

Fig. 57.—DIFFERENT INSECTS IN AN ANT-HILL (*Termes lucifugum*).
1. Workman. 2. Soldier. 3. Larva. 4. Nymph with small wings.
5. Nymph with long wings. (All magnified.)

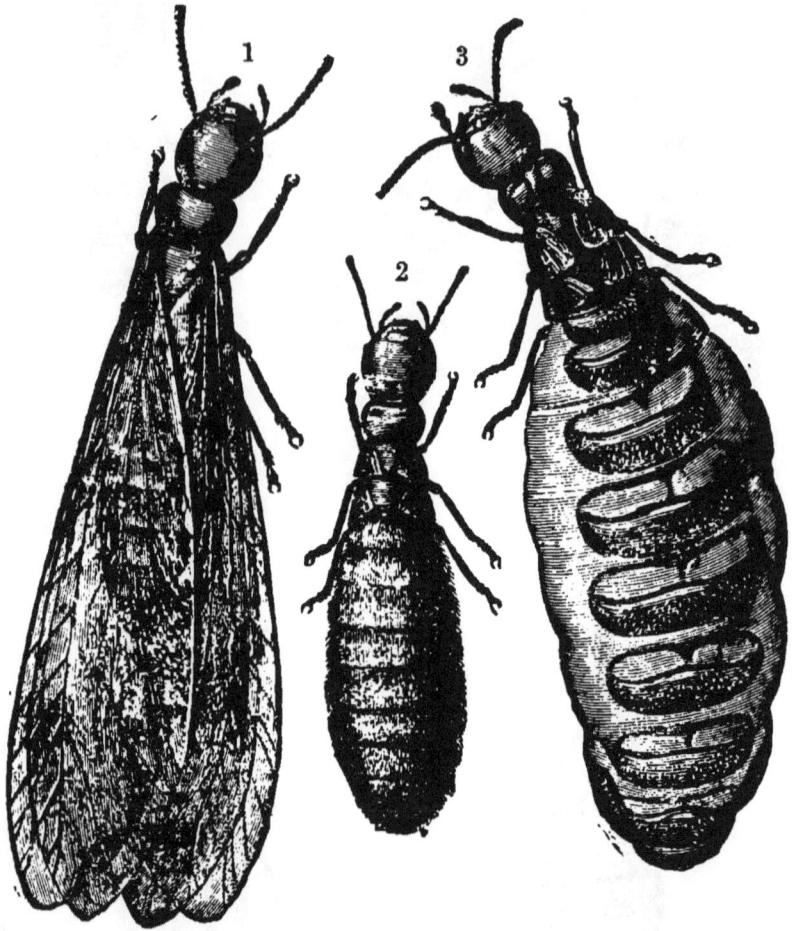

Fig. 58.—DIFFERENT INSECTS IN AN ANT-HILL (*Termes lucifugum*).

1. Male. 2. Little female. 3. Large female. (All magnified.)

and female meet each other, they cast off their wings just as the ants of our country do, and become the founders of a new colony. The soldiers are undeveloped males, and the workers are undeveloped females.

"The process of founding a new empire is very curious. Workers take charge of the king and queen and enclose them in a chamber which forms the first room of their many-chambered nest. They are careful to make the opening or door of this royal chamber too small for the occupants to pass through. This makes it necessary for the laborers to take entire charge of the eggs. For these they build small nurseries, at first very close to and around the royal chamber. When the nest becomes large by the increase of the inhabitants of the ant-city, these nurseries are made larger and at a greater distance. In the process of time their wonderful home is constructed, consisting of rooms for the royal pair, for nurseries and for the safekeeping of their provisions. These apartments are all connected by passage-ways filled with laborers and soldiers zealously pursuing their respective callings. The dwellings of the *Termes bellicosus,* the white ant of Africa, are from *nine* to *twelve* feet in height, and are flanked with little towers, and they are so solidly built that several men may stand upon them with safety.

"Here is a cut that gives a section of an ant-hill of Africa, and shows its internal structure." (Fig. 59.)

"Why are they called *termites ?*" asked Mary.

"The word *termes,* which is the name of a single ant, is Latin, and means a *branch* of a tree. I have no doubt this name was given to this insect because of its fondness for making nests on branches of trees, as is represented in the cut. (Fig. 60.) Men of sci-

Fig. 59.—NEST OF THE WHITE ANT (*Termes bellicosus*).

ence are sometimes governed by strange fancies in choosing names for families of insects.

"Here is a picture of a queen-termes distended with eggs. (Fig. 61.) Her abdomen becomes two thousand times as large as the rest of her body. She is then six inches long, and weighs as much as thirty thousand workers or as one thousand kings. She lays sixty eggs in a minute and more than eighty thousand in a day. This is not confined to a few days, but goes on at this r a t e throughout the year. T h e mother-bee does not produce as many eggs in a year as the queen

*Fig. 60.—*N**EST OF** W**HITE** A**NTS.**

of the *termites* does in a single day. If this family of insects had not a great many enemies who are ever on the watch to destroy them, they would soon become the masters of the world. Birds, poultry and the true ants devour them greedily. The inhabitants of Southern Africa are very fond of them. They roast them as we roast coffee, and then eat them by handfuls. Travelers also become fond of them, and compare their flavor to that of sugared cream."

"How strange it is," said Henry, "that men

should have a relish for such food as ants! It
must be because the people that live in Africa have

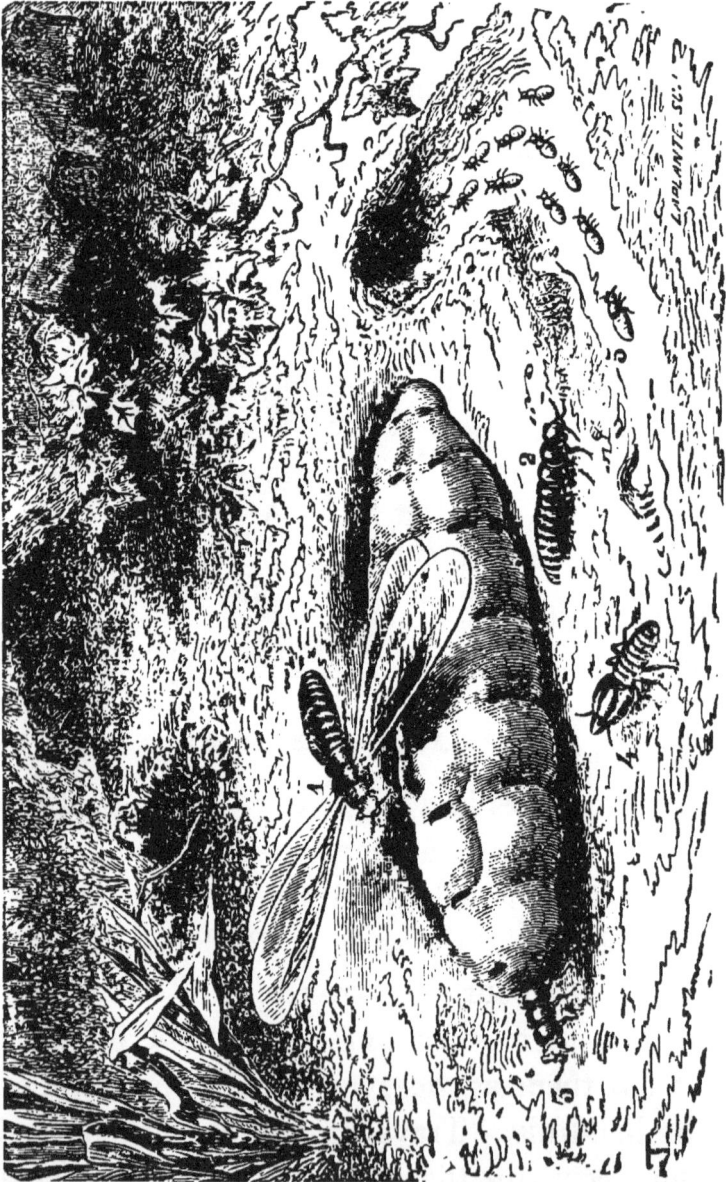

Fig. 61.—WHITE ANTS (Termes bellicosus) in Central Africa, after Smeathman.
(1.) The male termes. (2, 4, 5.) The neuters. (3.) The white ant-queen, distended with eggs.

scarcely anything to live on, and they are glad even
to feed on ants."

"Why, John the Baptist ate locusts and wild

honey in the wilderness," said James. "Do you think his taste for food was much more refined than the poor ant-eater's?"

"It is not long since," said Charlie, "that you both were full of your praises of the deliciousness of a dish of frogs, which you ate with a relish not certainly excelled by any of the ant-eaters. This matter of taste as to what we eat and drink depends much on climate and education. The refined African wonders as much at the barbarous tastes of Americans and Englishmen. I remember reading a story in Livingstone's *Travels in South Africa*, in which an African chief said to Livingstone, 'Did you ever taste white ants?' When the great traveler answered in the negative, he said, 'Well, if you had, you never could have desired to eat anything better.' The chief considered his own taste for food the standard for the whole world. So you are disposed to make your tastes the standard by which to judge the tastes of Africans.

"While I let the ant-eaters enjoy their dinner of termites, I see in their relish for that kind of food clear evidence of the wisdom of God in thus making provision for limiting the bounds of their rapidly-increasing numbers. It would very much derange the order of things if one species of insects should occupy a whole continent, destroying everything before it. But if what uncle says is true, it would not be long before this insect would drive from Africa all other animals except birds and fishes, and the whole country would be a dreary wilderness. This terrible result is prevented by the fact that it is good for food, and supplies the wants of men and ant-eating insects."

"You are quite right," said Uncle Samuel, "in your view of the divine wisdom. We see many things in the government of God which we cannot understand, while we can scarcely help seeing the highest wisdom in the results. And here we may well note the fact that in the animal creation there is a series of checks, by which the various orders are kept within bounds as to number. One order lives by keeping another order from multiplying injuriously. Man is at the head, with wisdom and strength given him to control the numbers of the lords of the lower creatures."

Here the conversation for the evening came to a close, and they all retired to the family-room, where, in sincere gratitude for the blessings of the day, they knelt around the family altar and offered up devout thanksgiving to God and sought his pardon, and commended themselves to his keeping—a fitting close to the duties and pleasures of the day.

CHAPTER XV.

IT was a beautiful afternoon. The sky was clear and the sun shone in his beauty, not parching everything with intense heat, but sweetly alluring the lovers of Nature to ramble among the wild-flowers and over the green fields. It was a holiday, and our young naturalists, full of zeal to learn more of the grand lessons of the divine wisdom so clearly taught in the insect-world, proposed to take Uncle Samuel with them and visit some of the haunts of insects not yet observed by them.

"Yes, cheerfully," said Uncle Samuel when asked to become one of the company of explorers. " I am always ready to go out on such excursions. But I must not forget my budget of pictures. Hand me my portfolio; we may need to look at some of the drawings before we return."

They had not gone far when little Mary, always looking out for something new, discovered in a sandy spot a little funnel-shaped depression, so regular in its outline that she at once suspected that some insect had been at work there.

"Look'ee here, uncle," she said; "how did this hole get here? See, it is as round as a thimble, and just like a funnel with its mouth up."

There was an anxious and interested company of observers suddenly gathered about the spot where Mary stood, and various conjectures were volunteered as to the cause of the liliputian sand-pit. But all agreed that it did not *happen* there by any accident. That it was *made* was evident, because there were marks of design in its structure, simple as it was.

Seeking a closer examination, Bertha dropped upon her knees on the ground and looked carefully into the little pit. She was soon rewarded for her pains by catching sight of the outstretched mandibles of the occupant lying at the bottom of the pit. (Fig. 62.)

"I see its arms," she cried. "What ugly things they are! Its body is all covered up in the sand. Poor thing! let's dig it out."

"No, don't Bertha—you'll spoil the pit. Wait till uncle sees it," said Henry, who interposed just here for the sake of science and because he wanted to learn the history of the strange insect that had attracted their attention, before it was disturbed.

"You must make way for me, then," said Uncle Samuel, "that my eyes too may rest upon the object which has suddenly produced such an excitement. Ah! now I see. I am glad you have made this discovery; it will amply repay you for this afternoon's ramble. The occupant belongs to the same family of insects with the white ants. It has a history intensely interesting. The *arms* which Bertha saw are its mandibles or jaws stretched out to

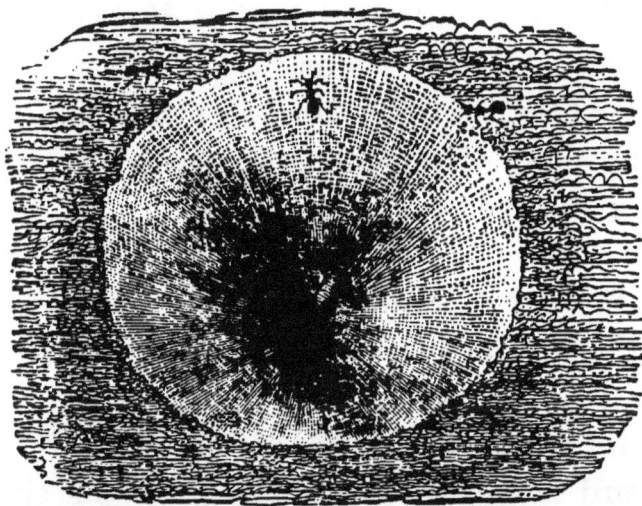

Fig. 62.—THE ANT-LION IN ITS PIT.

capture ants or other insects that may fall down the steep sides of its pit. Because it lies in wait for its prey in its own den it is called a 'lion,' and because it feeds chiefly on ants it is called the 'ant-lion.' It makes its sand-pit in its grub state; when it is a perfect insect it has beautiful wings and looks very much like a small snake-feeder. Its scientific name is *Myrmeleo formicarius.*

"In the engraving which I have here you can see the insect in three of its stages. The figure at the right shows it in its larva stage, that at the left in its pupa state, and below as it is when wrapped in its cocoon. But you can see the grub itself if you only blow two or three times into the sand-pit sharply, and bring out the insect to discover what is the matter."

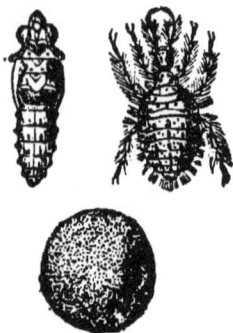

Fig. 63.—Larva, Pupa and Cocoon of the Ant-Lion.

This was soon done by James, and the sight of the lubberly insect increased the interest of the young observers in its singular history.

"You notice," continued Uncle Samuel, "that like all other insects it has six legs. You are astonished at its sudden disappearance and at the ease with which it buries itself in the sand. This backward motion is natural to it. Of its six legs, the hinder pair are the only ones employed by it in walking, and these, instead of moving the animal forward, drag it backward. When it takes a journey, its body sinks under the sand and its hind legs draw it back, so that its course is easily traced by the little hillocks it leaves behind it. It may well be said of it that it raises a great dust in the world when it travels."

"If this insect lives on ants and always walks backward, I'd like to know how it can capture its prey," said James. "It must have a hard time getting through the world with such awkward, back-handed legs as it has."

"I do not wonder at your question," answered Uncle Samuel. "It does seem impossible for it to make a living with all its disadvantages of locomotion. But the kind Creator never gave being even to the smallest insect without endowing it with such wisdom as is necessary to enable it to obtain its own food and to resist or escape its enemies; so he has not neglected the ant-lion and left it to starve, even if its food be one of the most active of insects. It has a lifework appointed for it as certainly as you have. It has also a higher life awaiting it, which it will certainly attain. And surely if such an uncouth grub as this can provide for all its wants and fulfill the end of its existence, none of you need ever despair of success in life. But you want to know how it supplies itself with ants enough to satisfy its wants. Certainly not by catching them by running after them, so God has taught it to devise a plan by which they come to it. This is an exceedingly curious device which no one but an ant-lion would ever think of.

"When it resolves to set up for itself and commence housekeeping, it selects a spot where the sand is loose and deep. Pushing itself beneath the surface so as to leave exposed its head and its jaws,

14

it begins the work of excavation. It shuts its long jaws, forming them into a kind of shovel, the sharp edges of which it thrusts sideways into the sand on each side of its head, thus lodging a quantity of the sand on the head as well as the jaws; then by a jerking motion of the head the sand is thrown behind its back or on each side. In this way it penetrates still deeper into the sand, throwing it away as it descends, until the pit has assumed the funnel-shaped appearance you witness. When the pit is sufficiently deep for its purposes it withdraws its head beneath the sand and leaves its jaws only exposed, and these are so spread out on the sand as to be scarcely visible. Now its 'dug-out' is ready for use, and the insect remains covered up at the bottom, calmly awaiting the fortune that sends its prey tumbling down the treacherous sides of the sand-pit into its open jaws.

"But here comes a little ant running along on its daily round of duty or pleasure, not suspecting that danger is near. Let us mark how this cunningly-devised scheme will work. See, it is approaching the edge of the sand-pit; now it halts to make observations, creeping rather cautiously still closer to the edge, as though to learn the true nature of the depression. It may be a philosopher determined at all risks to find out the cause of this opening in its pathway. One more step, and one too many. It has touched the loose sand, and down, down, helplessly it falls into the extended jaws of the hungry ant-lion. What a commotion in this little sand-pit! The happy occupant is sucking the juices of its prey, and will soon hurl its withered carcass out of the den."

"How wonderful!" exclaimed Charlie, "but all perfectly natural and right. The ant-lion is securing its living in a legitimate way—all well enough for itself, but death to the unlucky ant. Its pit makes me think of the pits which are everywhere open along our walks, into which the young often fall and are destroyed. They are the gambling-saloons, billiard- and drinking-saloons, card-tables and the theatre; evil companions are the loose sands, and the devil is the ant-lion. If a careless, heedless boy goes too near any of these pits, and evil companions get hold of him, he is almost sure to go rolling down the sides into the jaws of the great ant-lion, and the juices of his life are soon drawn out of him. For my part, I intend to keep far off from the edges of such pits, lest I slip as did this unfortunate ant, and my life become a failure."

"You have made the only safe resolution," answered Uncle Samuel, "and I am glad to know that the fate of the unwary and too venturesome ant has been so full of instruction to you. I hope you will all think how perilous it is to approach dangerous places. Keep far away from all precipices and all kinds of pits, and you will never be injured by any of them."

"But how does the ant-lion suck the juices of the

ant?" asked Henry. "It seems only to *hold* the ant with its long arm."

"Its mandibles are not solid, but tubular, so that it pierces the ant with them and sucks its juices. They are a fine mechanical contrivance, in which the wisdom of God is shown in adapting these instruments of capture to the higher wants of the insect."

"What could the ant-lion do if some larger insect than an ant should fall into his pit?" asked Bertha.

"Sometimes such a thing occurs, as when a spider carelessly walking about chances to step over the edge of the ant-lion's den. He is strong and able to contend for his life, and has no idea of permitting it to be sucked out of him without a struggle. The ant-lion soon perceives that his new prey is not to be as easily secured as the weak and helpless ant. The little quiet insect-home is soon changed into the scene of a desperate conflict. The spider makes every effort to escape, and the ant-lion, fearing lest he lose his savory meal, begins to twist his head about, shoveling up the sand with great activity, making the hole much deeper and its sides steeper, and causing such a disturbance in its sides that the spider finds it impossible, with all its great strength, to make its escape; at last it falls to the bottom, where it is soon disabled, and is held fast by the jaws of the lordly owner of the den till he has feasted upon its delicious juices.

" When you look at this lubberly thing crawling about the mouth of the sand-pit, and then at the neat, sprightly, narrow-winged insect in the picture (Fig. 64), you may wonder how such a remarkable change ever took place. And yet there is no miracle here. You see the animal in the two extreme stages of its existence. There is a stage of its life between these two in which it presents an aspect differing from

*Fig. 64.—*A*NT*-L*ION* (*Myrmeleon formicarius*).

both. This is its pupa state, during which it remains buried in the sand awaiting its transformation into its last and perfect stage, in which it shall be adorned as a thing of beauty and fitted for the full enjoyment of its higher-life state.

" Such a wonderful change in the shape and state of the same animal would scarcely be believed were it not established by accurate observation. And surely when the great Creator has endowed this diminutive insect with such astonishing life-power we ought not to doubt his own word when he tells us that beyond the grave there is a higher and better life awaiting man—a life in which he shall be gifted with purer and nobler tastes, holier desires and powers of movement far surpassing those he now possesses.

" I have no doubt that if there were philosophers among the grubs of the ant-lion, as there are among men, and it were told them that, after they had ceased to live as grubs and had wrapped themselves up in their silken cocoons to sleep their last sleep beneath the sand, they should rise again out of their silent tomb to dwell in the air with beauteous dress and gauze-like wings—that they should in their higher state of being no longer move in slow measures, crab-like, backward, and be compelled to wrestle with difficulties to secure the coarse food which the misfortune of other insects brought to their door, but that, eagle-like, swiftly darting through the air, they would eat more delicious food, and in that higher life breathe a purer atmosphere and mingle with insects of noble mien and gorgeous dress, and gaze upon the sun and reflect his glory in their seraph wings,—these students of Nature, ' wise in their own conceit,' would reject such teachings as the wildest ravings of the fancy.

" They would, no doubt, proceed to show how and why there could be no future state, no higher life—that ant-lions were made simply to spend their life in the bottom of their conical nest, draining the juices of the unfortunate insects that should fall a prey to their ingenious snare, and that when they had eaten their last ant and their physical powers were worn out they would pass away into their original nothingness or be absorbed into the great absolute insect. But how the sprightly, joyous, winged insect that has risen out of the abandoned shell of the buried ant-lion, with its new-formed body and loftier views of life, would laugh at the silly talk of the self-deceived philosopher, who tried to draw him away from his faith in a future life during his grubhood!"

It was an interesting scene—that group—during Uncle Samuel's long and somewhat profound talk; for, though the little girls did not quite understand all that was said about the ant-lion philosophers, still they were convinced by the earnest tone and the sparkling eyes of their uncle that the subject was intensely interesting, and they listened with fixed attention. The boys comprehended all that was said, and felt drawn to the study of insect-life with a new attraction as they listened to the arguments of their uncle. The philosophic Charlie was not slow to perceive the wisdom of God as displayed in the structure and habits of the ant-lion, and, having listened with unusual interest to his uncle, ventured to give expression to his feelings:

"Your account of this ant-lion, uncle, is truly astonishing. The more I learn of the wonders of the insect-world the more clearly do I see the goodness of God. When I see this pit with its tiger-like proprietor, and reflect on the fitness here displayed of the means to the end, I am filled with wonder and praise. I wonder at the curious expedient by which Nature has made up for the want of perfect organs of locomotion in the insect, and the remarkable and intelligent actings of that instinct with which it has been endowed, and I am filled with praise of the benevolent Creator whose greatness and glory do not prevent him from providing for the necessities of such apparently useless animals as ant-lions."

"Your thoughts might rise still higher," said his uncle. "You are each of more value than myriads of ant-lions; and I know that He who makes such wonderful provision for the grub-life of the ant-lion, and then lifts it up out of the grave and makes it a gorgeous and beautiful thing of elegant proportions and sylph-like wings, giving it a higher and better life, will not forget you, but will provide you with true soul-food—the truth of God—and after death will lift you up out of your graves and give you a nobler and holier life—a life of glory and immortality in the future world.

"One thing more I want to tell you about the ant-lion. When it commences to make its den it does not know what obstacles it may have to encounter. But it does not borrow trouble, as many men do, and neglect present duty, lest difficulties which it cannot master *may* present themselves. With unflinching courage it takes hold of its work, and labors perseveringly till it has completed it.

Sometimes a pebble too large to be tossed out of the pit by it is met. With earnestness of purpose that knows no discouragement, it takes up the heavy pebble on its head or back, and carefully walks up the side of the pit and throws it over the margin. Some observers say that to prevent its ever falling back into its pit it pushes the stone to a considerable distance from the mouth. If the stone is round and falls off the head or back of the struggling insect while ascending the treacherous side, it repeats the trial, until at last it overcomes every obstacle, and has the pleasure of taking its position at the bottom of the pit, cordially to welcome, and *feelingly* to embrace, every traveling insect that visits its mansion.

" You should all learn a lesson from the hero of my story. It is by moral courage, diligence and perseverance that you are to be successful in life. In every undertaking there are obstacles to success that must be removed. These are often very great, and will require trial after trial before success is reached. You know the song,

'If at first you don't succeed,
    Try, try again.'

Follow the example of the ant-lion, who gave not up the great object of its life because the first effort was not successful, but persevered till it conquered. If your lessons are hard, study them the more diligently and think the more intensely. If you are entering upon the responsibilities of life, and meet with discouragements in the way, face them courageously. Persevere. One by one your difficulties will be overcome, and in the end you will wear the laurel. Never forget the ant-lion and its lessons."

Here Uncle Samuel's interview with the young naturalists ended. An engagement to meet a gentleman on important business called him away, but not before he had promised to take another ramble with them whenever their school duties would permit.

CHAPTER XVI.

DURING the summer vacation our young friends, whose interest in the study of the forms and habits of insects was so gratifying to Uncle Samuel, spent much of their time in collecting specimens. Their zeal as students of natural history was greatly increased by their uncle's familiar talk about every capture that was new to them. While they were thus rapidly increasing their collection they were becoming familiar with the names and instincts of a large number of the inhabitants of the insect-world. It was a study of which they never became weary. Every walk in the fields brought them in contact with old acquaintances whose history they knew

or with strange forms that excited afresh the spirit
of inquiry.

"A new insect! a new insect! Let us go for
him quick!" At such a call all earnestly plied
their nets, till some one would cry out, "I've
caught him! he's my prisoner;" and then no time
was lost in finding their way to their uncle's study,
where they would soon learn its name and history,
and something about its mission in the world.

The example of our young friends is worthy of
imitation. There is no more exciting exercise than
a butterfly-chase, and the collection of insects
made during the summer would afford profitable
and interesting study for the entire winter season.

Now, it happened that one evening Henry caught
a large moth with brown wings, marked with an
eye-spot on each wing, the more prominent one
being on each hinder wing, and having two feather-
like antennæ on its head. (Fig. 65.)

"What can it be?" said Mary. "What big
wings it has! I wonder if it can see out of the
eyes it has on its wings? Let's take it to Uncle
Samuel."

Henry was very proud of his capture, and carrying it carefully he soon found his way into his
uncle's study. "I've caught him! I've caught
him!" he exclaimed as he presented his prisoner
to Uncle Samuel. "I've caught the king of all
the insects. Isn't he one of the giants? He
didn't care, either, for he let me come right up to

Fig. 65.—*Telea polyphemus*, sometimes called *Attacus polyphemus*.

him and put my cap over him; and then I put my hand under my cap, and he was mine."

"It must have lived a long time to grow so big," said Bertha. "Look at its two nice little wings growing out of its head! Aren't they beauties? I wonder if it flies with them?"

"You have indeed caught a very fine moth, and I do not wonder at your astonishment at its size; but you must not judge of the age of a moth, Bertha, by its bigness. So now, if you will all sit down and listen to me, I will tell you something about this splendid moth, and also about one of its relatives that has become very famous in the world."

"That we'll do with all our heart," said James.

So they seated themselves as near to their uncle as they could, Mary and Bertha drawing their little chairs up so close that they rested their arms on his knees. It was a happy, earnest group of listeners, with their eyes fixed on their uncle.

"This fine large moth is the *Telea polyphemus*; it belongs to the silk-spinning family, whose name is *Bombycidæ*. This high-sounding name comes from the Greek word *bombyx*, which means silkworm. The caterpillar of this moth spins for itself a cocoon of very fine silken threads, which have been manufactured into durable silk fabrics. It feeds on the oak, elm, hickory, bass-wood, walnut, butternut and thorn. I have found its chrysalis on the privet. It comes to its full growth in August or September.

The color of the caterpillar is a clear bluish-green.
Its head and feet are brown. Its warts are as lus-
trous as pearl, and are orange, rose-red and purple.
On the side of each segment is an oblique white line.
At its maturity it is more than three inches long
while feeding, but when in repose it hunches up
the rings, clinging to the twig with its back downward." (Fig. 66.)

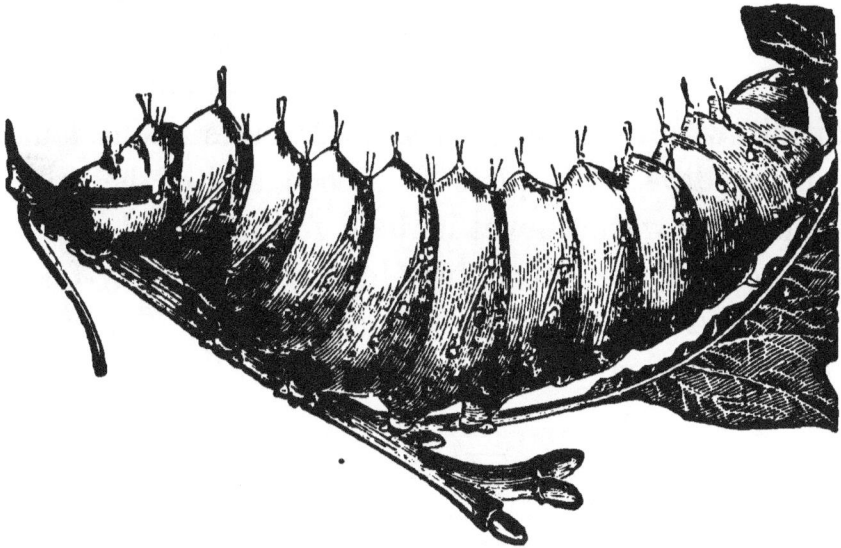

Fig. 66.—LARVA OF *Telea polyphemus.*

"How does it know when to spin its cocoon?"
asked Mary. "And has it any one to show it
how?"

"As soon as it is full grown it loses its appetite
for food and becomes restless, moving about as if
in search of some place where it can rest safely.
I have seen the worm stop and raise its head, moving it from side to side, as if it was taking a view

of its surroundings, and then it would remain still for a while, as if thinking of the wonderful change that had come over its tastes and plans for the future. It will sometimes commence its work of cocoon-building, and then abandon the place chosen for another. When it has finally settled on the locality in which it purposes to make its cocoon, it draws two or three leaves together very ingeniously by threads of silk as the foundation of it. It then begins to spin its covering, moving its head from side to side in zigzag lines, until it has surrounded itself with its silken shroud. It takes about four or five days to complete its cocoon, and in the operation the worm moves its head from one end of the cocoon to the other two hundred and fifty-four thousand times, never stopping to rest. Here is a good drawing of this cocoon. (Fig. 67.)

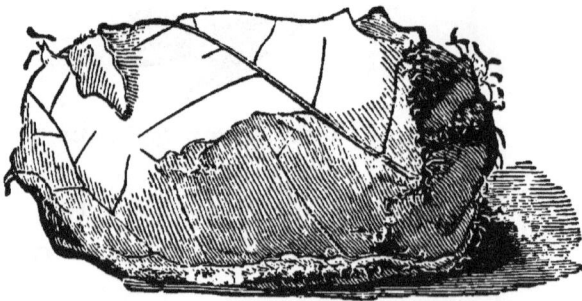

*Fig. 67.—*Cocoon of *Telea polyphemus.*

" When the work of cocoon-spinning is ended, it changes into its pupa state, in which it has an appearance very different from its larva state, as shown in this drawing. (Fig. 68.)

"This moth belongs to the genus *Telea*, group *Attacinæ*, which represents the largest kind of *Bombycidæ*. There are eight American species of the *Attacinæ*—the *Columbia, Luna, Splendida, Promethea, Angulifera, Cecropia, Californica* and the *Polyphemus*. The *Luna* is a very beautiful moth, and derives its name from the Latin word for *moon;* hence it is poetically called 'fair empress of the night.' The *Promethea* is called after one of the fabled Titans of the olden times, all of whom were of remark- able size. The *Cecropia*

Fig. 68.—PUPA OR CHRYSALIS OF *Telea polyphemus.*

derives its name from the ancient city of Athens. This is the largest of these moths, its wings expanding from seven to nine inches. The last one of this group, the *Polyphemus,* is the one which you have caught.

"This beautiful moth bears the name of one of the giants of mythology. *Polyphemus* was said to be king of the Cyclops, a race of fabled monsters who dwelt on the western shore of Sicily and lived on the flesh of men. They were of great stature, and had only one eye, which was situated in the middle of the forehead. I suppose the single eye on each wing suggested the name to the imaginative naturalist. Now, if you will only look up 'Polyphemus' in some good encyclopædia or work on ancient mythology, you will find out many more things about this man of one eye, and how Ulysses crippled him, and then cunningly made his escape out of his cave."

"I have read all about the Cyclops and their king," said Charlie, "but I never supposed that their history would be recalled to my memory by a moth. But I am anxious to hear about the relative—cousin, I suppose—of this moth, which you say has become so famous all over the world."

"This relative is not as large or as handsome a moth as this one. You would never think of calling it 'the king of all the insects,' and yet it is more worthy of the honor. While our *Polyphemus* moth, with all its magnificence, is but little known and is noticed by few, its relative, of moderate size and unassuming in its dress, is reverenced by kings and gives riches to nations. Though unadorned by any brilliant spots or lines of beauty, queens are indebted to it for the richest robes that decorate their persons. It is about one inch long, and its wings extend about two inches. The antennæ of the males are small and feather-like; they fly about in the evening, and sometimes by day, but the females are inactive. These lay their eggs on the mulberry tree, on the leaves of which the caterpillar feeds. The eggs are about the size of a mustard-seed, and the worm is very small when it is hatched. Its appetite is capital, and, as it eats ravenously, it grows in a very short time three inches long, or nearly five hundred times as large as it was when it first saw the light. If either one of you should grow as large in proportion, when full grown you would be more than four hundred and fifteen feet high.

" This is the worm that produces the fine silk of which our dresses, shawls and ribbons are made. Its name is *Bombyx mori.* You know that *bombyx* means silkworm, and *mori* means of mulberry, from the Latin word *morus,* so that the name when translated means '*mulberry* silkworm.' This worm has a remarkable history. Its ancestors lived in the world thousands of years before their value was known. All this time they dwelt in the woods, spinning away at their silk, but without attracting any attention. At length, as the story goes, a long time before Christ was born, one of the queens of China, while walking through the royal grove, noticed one of the cocoons made by this worm, and, plucking it from its fastenings, she took it to the palace. Examining the threads of which it is composed, she admired them for their fine quality, and discovered that she could separate them and wind them in a ball for use. She then collected a great number of cocoons and wove the threads into fine cloth for wearing apparel.

" This was a very important discovery, because up to that time the clothing of the Chinese was made from the skins of animals, and there were not enough skins to supply the wants of the increasing population of that great country. There was, therefore, a great demand for some other fabric out of which good clothing could be made, and the silk of this newly-discovered worm was so employed. It was a great thing for this neglected creature, that had been kept away from the habitations of man for centuries, to be raised at once to the rank of a favorite of queens and royal ladies. But so it was.

"I do not assert that this is a true story, but the Chinese believe it, and they say that the name of this queen was *Si-ling-chi*. Now, if ever such a woman lived it is well to remember her name, and the Chinese have done well to tell us who she was. Her husband, the story says, built an enclosure joining his palace, where the mulberry was planted and the silkworm was raised; so every day the queen went into that enclosure with her maids of honor to feed the worms and gather the cocoons, and the finest pieces of silk were always woven by her own hands. Si-ling-chi was an industrious queen, and the Chinese may well regard her as one of the best of their race.

"It was not many years till the people in China made large quantities of silk and sold it to other countries. It became, therefore, a source of great wealth to the Chinese, and for fear other nations would learn to make silk for themselves the emperors forbade the carrying of the eggs of the silkworm out of the country. In this way for a long time the manufacture of silk was confined to China, and its price was greatly increased. About the year 300 before Christ, when Alexander the Great was but a little boy, silk was sold in the cities of Greece for its own weight in gold.

"Julius Cæsar, who was born one hundred and two years before Christ, was the first who introduced silk to the Romans. He dressed himself in silk robes. After his death the emperors and rich men became very extravagant in the expensiveness of their silken garments; but they did not know how the silk was made, and the people had no idea that the splendid robes of the emperor and the senators were the product of a worm. They supposed the material to be of vegetable origin. Aristotle and Pliny, two learned naturalists—one of Greece and the other of Rome—said that a caterpillar made the silk, but they were not believed.

"I know you will be glad to learn that some one succeeded in getting the eggs of the silkworm out of China and to introduce the manufacture of silk into other parts of the world. It happened in the year 530, when Justinian I. was emperor of Constantinople, that two monks were sent on a mission to China. While there they learned how to raise the silkworm and to manufacture silk. Securing a number of silkworms' eggs in their hollow canes, they returned to Constantinople. From this small beginning the culture of silk took its rise in Turkey.

"Spreading into Greece, the cultivation of the mulberry became so general there that that peninsula was called *Morea*—that is, *Mulberry*—peninsula. It was also noticed that the shape of the peninsula was that of a mulberry-leaf, as though it had been originally designed for the home of the silkworm. A long time after this, in the year 1340, the people of France began to raise the silkworm and to manufacture silk. This trade has now become a very important one in that country. England did not manufacture silk successfully till about the year 1824. Many attempts have been made to raise silk in this country, and with little success; but the *manufacture* of silk is carried on with decided success in the States of Connecticut and New Jersey.

"Here is an engraving which represents the insect in all its stages. You see two caterpillars, full grown—one just getting ready to feed on a leaf, and another commencing to spin its cocoon. A cocoon is represented as laid open, and the chrysalis lies on the branch below, while the male and female moths spread their wings at the top of the picture.

"When the caterpillar has arrived at its full growth and the days of its wormhood are numbered, it looks out for a good locality in which to spin its cocoon. It calls not upon its insect-neighbors to make its shroud and lay it in its tomb, but proceeds in a business way to weave its own grave-clothes out of material of its own spinning, and to lay itself down to rest and wait till its great change shall be complete. It takes it about three days to make its

Fig. 69.—THE SILKWORM IN ALL STAGES OF ITS EXISTENCE.

cocoon, during which it makes no less than three hundred thousand movements, or a little more than one every second.

"This part of its work being done, it gradually changes into a brown-red chrysalis. From fifteen to seventeen days the little worm seems to rest within its chrysalis-case. But it is not resting. There are remarkable changes going on within this dark workshop of Nature, which are the result of a wonderful activity. The entire form of the caterpillar undergoes a complete change. Thousands of distinct eyes, all acting together, are set on each side of its head. Feather-like horns are placed on the head near the eyes. Wings attach themselves to the sides of the thorax, and these are covered with scales which serve for ornament as well as protection. The pro-legs of the caterpillar are rejected. and the six legs of the insect assume proportions and a length adapted to the new life on which it is destined to enter. When all the apparatus of life needed in its new state of existence is complete, and its instinct moves it to seek release from its imprisonment, it splits open the dry casing in which it is confined, and is ready to make its way through the outer silken wall of its prison, that it may enjoy the freedom of its higher life."

"But how can the poor thing ever make its way through such a covering of silk? It seems to me it would certainly perish in the attempt," said James.

"It can easily overcome that difficulty," said Uncle Samuel. "It has a little vessel in its head containing a peculiar liquid, which, as soon as it has freed itself from the chrysalis, it throws out on the cocoon. The threads of silk are moistened by this liquid, but not broken. All it has then to do is to push aside the moistened threads and go forth into the daylight beyond.

"When it first stands upon its cocoon and takes a survey of its new situation, its wings are wet and folded back on themselves. But they do not long remain so. They begin immediately to expand and take their proper shape, and soon the silkworm, with its new organs and new tastes, is ready to perform the duties and drink in the pleasures of its new life.

"Before laying her eggs the female selects with care a locality in which they will be safe and where the young will find abundance of food. She then lays them side by side, covering each with a liquid which glues it to its place. She lays from three hundred to six hundred eggs. When they are all laid, which usually takes about three days, having fulfilled her mission and provided for her offspring, she dies. When the little worm is ready to leave the egg it gnaws a hole through the shell and escapes. It is from the first provided with apparatus for cutting the leaf upon which it feeds, and it loses no time in putting this apparatus to a good use."

During this unusually long talk our young naturalists were intensely interested. They knew that the silk bought in the family came from the other side of the ocean, but they never before knew the history of the wonderful little worm that manufactured the material out of its own substance. The little girls were specially interested, and wondered if the worms knew they were spinning silk to make ribbons and dresses for them. Charlie listened thoughtfully, and as his uncle progressed in his story his thoughts took shape and he now spoke for the first time.

"I wonder," said he, "of what use the silkworm would have been in the world if man had never been created ? No other animal could have used the fine silk of its cocoon, and, as the result proves, it was certainly intended for use beyond the simple protection of the chrysalis. It seems to me that God designed the material of the cocoon to meet the wants of man, and for this reason he gave skill to the worm to manufacture the silk, and wisdom to man to discover its usefulness and the genius to work it up into garments and other articles necessary to his comfort. Does not the fitness of the silkworm to the wants of man prove unity of design in the creation and structure of each ?"

"Your reflections are very good and your question pertinent, and it certainly admits of but one answer, and that an affirmative one. It is very strange that any reflecting mind should fail to see the unity of design, so manifest in the different departments of nature—how one animal is made to minister to the welfare of another, and how one part of nature depends on another for its continued existence and comfort. You study natural history profitably if you notice carefully the evidence it affords of the wisdom and goodness of God. All his works praise him, and should not we?"

This long talk exhausted the evening hours, and was a fitting preparation for the family religious services to which they were called just now. The scene changed from the study to the family-room, but the latter presented a no less interested group of listeners when Uncle Samuel read the Scriptures and led in prayer for a blessing on the household for the night.

CHAPTER XVII.

"LOOK here, quick, uncle! See this splendid spider and its pretty home. Its fine large net is as evenly knit as Charlie's fish-net, and it sits in the middle of it as proud as a king. Here it is on this lilac-bush. Just look at the old spider! Isn't he independent, though? But what an ugly-looking insect he is! I would rather hold a great hairy caterpillar in my hands than him."

When the impulsive little Mary thus addressed her uncle he was walking in the garden indulging in one of his meditative moods on a pleasant afternoon in midsummer, not observing that the children were also there, looking about among the bushes and vegetables for objects of interest. Looking around, he saw the little group gazing intently upon the home of a spider, which remained in its position all unconscious of being the object of attraction to so many intelligent eyes.

"You are mistaken, Mary," said Uncle Samuel, "when you call the spider an *insect*. But you can

Fig. 70.—THE SPIDER AT HOME.

be excused, since some naturalists have classed it among the insects. If you count its legs, you will find that it has eight, while insects have only six legs, and the spider has no antennæ, as insects have. Nor are its eyes, like the eyes of a moth, made up of a great many small eyes united in one, but they are simple, and vary so greatly in different spiders that they are divided into groups according to the kind of eyes they have. The number of the eyes of the spider is four, six, eight or ten, according to the group. These are all without eyelids, and are covered with a hard, polished, transparent crust. Being immovable, they are multiplied, and so located as to enable the spider to see in all directions.

"They belong to the same general division of the animal world with insects—the articulates—but they are only a kind of thirty-second cousins to them. Now, you know that such a distant relationship as that is not worth minding. So they are set off by naturalists as part of a family which includes scorpions and mites—a rather disreputable brotherhood. This family is called *Arachnidæ.* The *true* spiders form the section *Araneidæ.* You must not, however, despise spiders because they have bad relations, for with all their faults you will find in them very much to admire and esteem."

"Tell us all about spiders, uncle," said Henry. "I never thought of their not being insects. Mother often says when she sweeps one down from the ceiling, 'The ugly insect! kill it; it always puts itself where it's not wanted.' How do they spin their fine silk? and what spider is this one which has made for itself such a beautiful web?"

"Come, then, let us sit down in the summerhouse; and, James, please go to my study and bring my portfolio; I will need some drawings it contains. Our position will permit us to watch the spider and see what it does when a fly becomes entangled in its net."

As soon as they were seated and the portfolio was brought, Mary seated herself close to her uncle, for she felt that the spider-talk belonged specially to herself as the discoverer of the spider-net.

"You will want to know how the spider came to have such a learned family name. According to an ancient story, there once lived in Lydia, a country in Asia Minor, a beautiful lady who was a very skillful spinner, and whose name was *Arachné*. She was a very ambitious young lady, and believed that she could spin faster and better than Minerva, the goddess of wisdom; so Miss Arachné challenged the goddess to a spinning-match. The challenge was accepted, and when Minerva saw that the young lady of Lydia was about to gain the day, she struck her on the forehead with a spoke of her wheel. This so insulted Arachné that she went away and hanged herself, and the goddess, repenting of her rashness, kindly turned the young lady into a spider, so that she might spin for ever; so the story goes. So the spiders have been very great spinners ever since, and they keep up the custom still of hanging themselves by a thread when insulted, as perhaps you have seen the garden-spiders do. Hence they are called *Arachnidæ* in Greek and *Araneidæ* in Latin."

"Yes," said Bertha, "the thread seems to come

right out of the spider. I wonder where it gets the thread? Does it make it, uncle?"

"The 'spinning-machine' of the spider is so wonderful and yet so simple that I must describe it to you. If you examine the hinder extremity of the abdomen of the common house-spider (*Aranea domestica*), you will find on its under side four protuberances of a cylindrical shape, which are called *spinnerets.* Each spinneret is furnished with tubes so very fine that a thousand of them are found in a space not much larger than the point of a pin. From each of these tubes proceeds a single strand, which unites with all the other strands to make one thread, so that the spider's thread, often so fine as to be almost invisible, is composed of *four thousand* strands. Naturalists say that some spiders have five, and others six, spinnerets, each of which has one thousand strands, making the resultant thread to be a rope of five or six thousand strands. The silk is produced from the food by a process going on in several silk-reservoirs within the animal, where it remains until it is needed.

"We admire the spinning process invented by Arkwright and perfected by others, but human skill has never produced machinery that rivals the spider's spinning-jenny. I have read of a manufacturing firm in Manchester, England, that produced cotton thread so fine that one pound of it would reach ten hundred and twenty-six miles, but it was too delicate to be handled or applied to any useful purpose. The spider exceeds this every day. While it spins a thread of much greater fineness, it applies it to the most useful purposes—the capturing of its prey and the building of its dwelling-houses and sailing vessels. Travelers tell us that in Java spiders' webs are met with so strong as to require to be cut through with a knife.

" But, uncle, I do not see why a spider must have so many strands in one thread, for all the use it puts it to," said Mary.

" There may be two reasons for this arrangement, Mary. First, the dividing the threads into so many strands favors their rapid drying, which is very necessary, because the spider wants to use his strong ropes just as soon as they are spun. Secondly, the union of so many threads in one greatly strengthens the cord, and it is necessary to the life of the spider that the threads of its net be strong enough to resist the force of the flight of a strong fly and hold it till it is captured.

" The only other instruments used in spinning are its feet, with the claws of which it guides or separates into two or more the line from behind. Here is a drawing of the spider's foot as seen through the microscope. (Fig. 71.) You observe that it is triple-clawed. One of the claws acts as a thumb, the other two being toothed like a comb for gliding along the lines. With these two claws it keeps the threads apart. When the spider ascends the line by which it has let itself down, it winds up

16

the cord into a ball. For this purpose it uses the third claw, which I have called its thumb."

"How did the word *cob-web* originate?" asked Charlie.

"The old Dutch word for spider was *coppe*, and from that word sprang the word *cob-web*, meaning spider-web. This word always suggests the idea of neglect or desertion, because spiders' webs are found in deserted places, being swept away where active life is going on. It is not regarded as an evidence of good housekeeping when spiders are permitted to retain possession of the corners of the ceilings. Hogarth, when he wished to represent charity as neglected, closed the charity-box, in one of his pictures, with a cob-web. There is a legend among the Jews that when David entered the cave of Adullam a spider wove a web across the entrance of the cave so quickly that Saul passed it by, convinced that David could not have entered it for refuge.

Fig. 71.—SPIDER'S FOOT.

"The spider whose net you have so much admired is called *Epeira diadema*. It is very frugal and industrious. It means to secure a living, and, in its way of thinking, an *honest* living. Is it not just as honest for the spider to trap insects as it is for boys and men to trap the hare and shoot the deer? Is it not more innocent than for boys to rob birds' nests? I think it is. You must not, therefore, condemn the spider because he captures the innocent fly. The beneficent Creator has given it an instinct by which it weaves a net for catching its prey, and has provided it with material for this purpose. In living as it does it is evidently carrying out the end of its being and keeping other insects in check, and should therefore be praised for its faithfulness, rather than condemned for its cruelty."

"I read somewhere," said Charlie, "that the spider makes its web with the exactness of mathematics, and that it is therefore called a geometrician. Does it make calculations, lay down a plan for weaving its web, and work by it?"

"The regularity of the spider's web, as you see, is very remarkable, considering that the builder had no scale for measurement and worked altogether by the eye, and perhaps never made a net before. It is in compliment to the evenness of its web that it is called a geometrician, and not because the lines are accurately mathematical.

"In the construction of its web the spider acts like a mathematician so far as to fix upon a centre of operations. From this centre a few lines of silk are drawn out and made fast to certain points which are to be the boundaries of the web. These points are next connected by a thread of silk which forms the outer edge of the web. The centre is then connected with this line by numerous threads of silk, like spokes of a wheel. Now the framework of the web is completed, and our ingenious weaver draws out the concentric circles which you see are connected with those straight lines that pass from the centre to the circumference. With these circles the entire web is filled up, and its work is done. If you examine the web closely, you will find that the spider has covered some of the threads with a kind of glue or paste, which clings to the feet and wings of any insect which may happen to fly against it."

"Isn't he cruel," said Bertha, "just to arrange his web so that the poor fly's feet will stick to it? I think he means to catch every fly he can. How wicked!"

"Not so wicked, after all. It is the spider's way of living, and if it is right for you to eat the flesh of the innocent turkey on your birthday, it is not wicked in the spider to catch the innocent fly and serve it up for its breakfast. So when the wonderful spinner has completed his fly-trap, he places himself in the centre, where he now is, and extending his feet so as to touch all the rays of the web, he quietly waits till some unwary insect caught in his snare rouses him to action."

"Just look!" said James; "there is a fly caught in his web. See how it struggles to get free! But the more it struggles the more it wraps the web about it. Now its wings are all covered, and the spider is striking it with its claws. Poor thing!"

"Its claws are poisonous," continued Uncle Samuel, "and hence its strokes are fatal to the fly. But the poison that kills the fly doesn't make it unfit for food for the spider. Already he is drinking the juices he draws out of the fly, and enjoys them as much as Bertha did the turkey."

"But suppose," said Mary, whose sympathies were now enlisted on behalf of the spider, "no fly should come in the way of the net all day, and all the next day and the next, what would the poor spider do for something to eat? Would he not starve if the flies should all agree to fly to some other place and leave the web alone?"

"This little animal," replied Uncle Samuel, "is so made that it can live many days without food and not suffer from hunger. This wise provision for the continued well-being and happiness of this apparently insignificant creature shows how much the Creator loves the works of his hands, and it ought to teach you to trust in him to sustain and provide for you, seeing that you are of more value than millions of spiders.

"I always pity the innocent fly that is so unfortunate as to take the direction across which such a net is placed, for I know there is no help for it. It must die a cruel death. Its enemy is stronger than it is, and has it in his power; and when I look at its vain struggles for liberty and life, I think of many unfortunate boys and girls who have been caught in more fatal snares and who are not as ignorant as the fly. You cannot teach the fly anything. It was not made to be taught. So when I see it entangled in the spider's web I say, 'Poor thing! it could not help it; it did not know any better.'

"But when boys are told that there are fatal soul-nets—webs of death—in every drinking-saloon, billiard-saloon and theatre in the land, and that if they visit these places for amusement they may be caught by a terrible enemy, worse than any spider, who will strike them with his poisonous claws, and they will not listen to the voice of warning or care what may become of their souls, but enter in and are ensnared and lost, I greatly pity them, but I say, 'What fools, to put themselves in the power of their greatest enemy when they knew better!'

"Now, my dear young friends, I want you to go around all these soul-traps which the devil has set for you. It may take you some time to get round them, and require some effort, but you will reach the true end of life sooner and more safely by doing so. If the poor fly that was just caught and devoured had only gone around the spider's web, it would have lengthened its flight, but it would have reached the end of its journey in safety. It took, however, what seemed to be the most direct and agreeable course, and it was caught by its foe; and that was the last of it.

" If you want to succeed in business, do not enter into an unlawful occupation, no matter how much money you can make at it; for if you do you will be caught in a soul-snare, and that will be the last of you. You may gain wealth and ease and world-ly honors, but you will lose your soul. Commence active life in an honest business, do your work well, never be above your business, never act against con-science, never go into bad company, and, although you may not seem to be prospering as some godless men do, you will form for yourselves a good reputa-tion, and in the end, if not wealthy, you will be respected and happy."

" I read a curious account of a spider found in the West Indies," said Charlie, " that makes its nest in the ground. It digs a hole in the earth obliquely downward about three inches in length and one in diameter. This cavity it lines with a tough, thick web, as tough as leather. This long deep home has a door hung on hinges, which is opened and shut whenever the occupant passes and repasses. Is not this a remarkable spider?"

" It certainly is. But there are many such spi-ders in the tropical regions; they are called *mason-spiders*, because they use sand and clay in construct-ing their nests. The spider whose nest you describe is called *Mygale nidulans*, and a full account of it may be found in Darwin's *Zoonomia*. It borrows its generic name from a Greek word, which means a *shrew-mouse,* and is so called because it burrows in the ground like its namesake.　Its name *nidulans* means *nest-digging;* so you have its name in English—the *nest-digging shrew-mouse.*"

" What a queer name for a spider!" said Bertha. " I think the people that live there don't know much when they call a spider a mouse."

" I have a picture in my portfolio of another

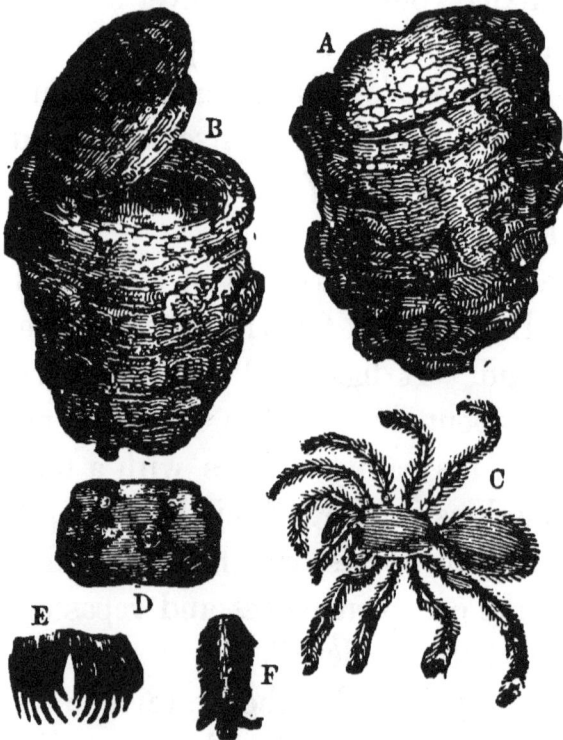

Fig. 72.—MASON-SPIDER AND NEST.

A, the nest shut; B, the nest open; C, *Mygale cementaria*; D, the eyes, magnified ; E, F, parts of the foot and claws, magnified.

spider of the same genus that makes its home in the south of France. (Fig. 72.) It name is *Mygale cementaria.* It is called *cementaria* because it is found in the region of the Cevennes Mountains—

the name of the mountains and that of the spider being derived from the same Greek root. It is very careful to select for its nest a place that is bare of grass, and that slopes so as to carry off the water. It digs a hole perpendicularly about two feet in depth, and makes it large enough for the inmates to move up and down easily. When the walls are prepared, it lines them with very fine silk, which it glues fast to their surface. It closes the opening with a door constructed by fastening several layers of earth together with silk. This door is flat and rough on the outside, corresponding in appearance with the ground around it, so as to conceal the nest. It is convex on the inside, and lined thickly with fine, strong silk. This lining on one side of the door is so attached to the rim of the entrance as to form an excellent hinge. The door, when pushed open, shuts with its own weight. The spider ordinarily remains at the bottom of its deep den, but is so sensitive at every movement of the door that if any insect or other intruder attempts to open the door, it ascends immediately, and seizing with its claws the silk lining of the door and clinging to the lining of the tube, it holds the door down with all its strength. If it is not strong enough to resist the entrance of the intruder, it retreats suddenly to the bottom of its den and awaits the result. It leaves its home at night to capture its prey, which it brings to its den to eat at its leisure."

"This spider," said Charlie, "shows a great deal

of cunning in making the outside of its door flat
and rough, so that no insect wandering near would
suspect that there was danger close by.   In this
provision one cannot help seeing evidence of intelligent design."

"Your conclusion is the only logical one that
could suggest itself to a thinker, and it is wonderful that any person can study the history of the
animal world and doubt the existence and attributes
of the great Jehovah.   But the device of this
cunning spider suggests similar devices employed
by the spirit of evil among men.   It has always
been its plan to conceal itself.   Infidelity sometimes
hides itself under the style and language of honesty,
and deceives by its very semblance to truth.   Satan
appears among men as an angel of light, and deceives
many.   So it happens that very many young, unsuspicious persons are caught in the trap that is ingeniously dug for them, and the door is shut down
on them, and they are either kept there always or
fight their way out with great difficulty.

"Evil companions are trap-door spiders.   They
are very polite, kind, accommodating, pliant, and
sometimes intelligent.   They will agree with you in
everything, go with you to church, talk religiously,
and even put on the appearance of great piety, if
all this is necessary to get you in their power.   But
this is only the outside finish of the trap-door that
conceals from your view a wretchedly wicked heart
and deep-laid schemes for your ruin.   Alas! how

many unwary and trustful souls have been deceived by such treacherous companions, and have become their unhappy victims for life. You are young and unsuspicious; you have not seen so much of the wickedness of the world as have we who are older. You must take our word for it that there are dangers against which you must be on your guard.

"Take warning from the story of the mason-spider. Think of the name *trap-door* spider which its concealed den has earned for it, and beware of the moral trap-door spiders which are concealed along your life-journey. You need a guide to point them out to you and to keep you from falling into their concealed pitfalls. God has given you such a guide in the person of his Son Jesus Christ. Only follow him and you will be kept from all harm, and your feet will never glide into any of these pits of moral evil."

Uncle Samuel never failed to impress his young friends with the moral lessons which the revelations of animal life suggested; and it was difficult to tell whether his eager pupils were more apt to learn the facts which he taught them or to practice the laws of life which he so forcibly illustrated by the habits of the insects with which he made them familiar. It is certain that religion had a stronger hold upon their young minds at the close of each conversation because of the light which was cast upon duty, and the evidence each insect afforded of the wisdom, kindness and faithfulness of God. At the conclusion of Uncle Samuel's story his little earnest crowd of listeners was dismissed to the croquet-ground to enjoy themselves a while before supper, while he continued his meditative walk among the flowers, glad at heart that he was able to add so materially to the enjoyment of his nephews and nieces, who were so eager to obtain knowledge.

CHAPTER XVIII.

"JUST see how it skims over the water faster
than a duck can swim," said Mary.

"There!" said James, "it dives under the water;
see, it is up again, giving a flirt in the air as if it
were shaking itself dry. It don't mind the water,
as that spider did that we put on a little paper-boat
in a basin of water yesterday. It would put out one
of its fore feet and touch the water, and draw it up
again, as if it were saying to itself, 'I must keep on
dry land; there is no walking on the water, that's

253

certain.' Then, you know, when we took the boat
from under it, it sank to the bottom, and when
taken out it was wet all over and could scarcely
walk with its load of water."

Our young naturalists were making observations
by the side of a pool in which some water-spiders
had made their homes, having become greatly interested in their novel and strange habits. They
had supposed that spiders were confined to the land,
and their experiment with a garden-spider had confirmed this belief. But now they saw something
which was to them very much like a miracle—a
real spider sporting in the water like a fish. Could
it be that it had its home there? They saw no
web nor anything that looked like a spider's house.
When the spider sprang up in the air it seemed to
them to be carrying a ball of glittering silver. It
did not remain long above the water, but would
dive back again and hide itself for a few moments
among the plants under the surface. What did it
do down there? How did it get air to breathe
under the water?

These questions our young naturalists could not answer satisfactorily. They conjectured many things,
but, like older philosophers who endeavor to account
for what is beyond their knowledge, they could not
prove any of their conjectures to be true.

It was very fortunate for them that just at this
time Uncle Samuel happened to pass that way while
taking his afternoon walk. I say *happened*, not

because it is the best word possible—for I do not believe that anything happens *by chance*—but because his appearance just then and there did seem so much like a fortunate occurrence that we cannot blame our young friends for thinking that he came, by some good luck, just at the very moment when they wanted him most.

"We are so glad you happened here just now, uncle!" said Bertha. "We've been watching a little spider in the water, and it's the 'cutest thing you ever saw. It cuts all kinds of pranks on the water, and dives under and stays down almost long enough to drown itself. But it don't drown; it comes right up again, and jumps up and shines as bright as silver. What does it do in the water? Does it live there? How can it? And how can it catch flies to eat when it has no web?"

Bertha's description of the habits of the spider was full enough to enable Uncle Samuel to recognize the species to which it belonged, and he was ready at once to tell them all about the wonderful air-castle it built for its home under the water.

"The spider you saw is a water-spider," said Uncle Samuel. "It would find as much difficulty in making a living on dry land as the beautiful garden-spider would find in making its living in the water. The good God has adapted each of his creatures to the kind of life it lives, and has richly provided for the comfort and happiness of all kinds of spiders.

"If the instinct of the spider guides it to seek its prey in deserted halls, or in kings' palaces, or among the garden shrubbery, he gives it a perfect spinning-apparatus and material with which to construct a web, as a net, to ensnare the unsuspecting fly, the dainty food on which it lives. If its instinct leads it to make its home in the ground, it is supplied with strong claws for burrowing and skill to conceal its burrow by means of a very ingenious trap-door, while it makes its little home tasteful and comfortable by lining it with a silken kind of felt that no animal but itself can make.

"If its food can be found only on the surface of the gently flowing stream or the sleeping lake, it is gifted with a talent for constructing just such a sail-boat as it needs, and mounted on its tiny leaf-vessel it goes forth watching for its prey and successfully capturing it. Or if it be a *pirate spider*, it needs no boat or raft, but glides upon the surface of the water in pursuit of its prey. As one has said of this species,

'They bathe unwet their oily forms, and dwell
With feet repulsive on the dimpling well.'

"If it is prompted to build an air-castle under the water, its instinct does not fail it, and its structure and its skill enable it to confine globules of air in the midst of the water, so that it can rest in calm security in its air-hall waiting for the approach of the insect on which it makes its breakfast.

"The spider which attracted your attention is one which builds its palace of air under the surface of the water. Scientific men have given it a very long and high-sounding name for a spider. They call it *Argyroneta aquatica.* This name means *water-nymph.*"

"What are nymphs, uncle?" asked Mary.

"Charlie has been studying ancient history, and ought to be able to tell you," answered Uncle Samuel.

"I think I can," said Charlie. "The ancient Greeks believed that all Nature was filled with spirits, and these spirits were sometimes called *nymphs.* So they had nymphs of the groves, of the valleys, and of the sea. They believed that the sea-nymphs lived in caves and grottoes on the coast; that at night they glided along the shore; that their hair hung down their backs in beautiful flowing tresses studded with coral and pearls. They said that when their prince, who was a son of Neptune the sea-god, would give a blast on his shell, they would all plunge into the blue water to attend the chariot in which the wife of Neptune rode out for her health. They are described as very handsome women, who always sat astride of dolphins when they took a ride on horseback. I suppose this spider is called a *nymph* because it dives into the water and makes its home there."

"I don't think a spider looks much like a handsome young lady," said Henry. "Naturalists must

17

have a strong imagination to fancy that such a diminutive thing as a spider was a god with flowing tresses, though I can see how this spider would suggest the idea that it was studded with pearls, for it glitters just like the most brilliant pearl when it leaps out of the water."

"Well now that you know its name, I must tell you how this curious little water-nymph builds its house in the water and supplies it with air; for its palace is a wonderful specimen of architecture, and reflects great honor upon the taste and skill of the aquatic animal as a practical philosopher. When it resolves to provide a home for itself, it first fixes upon the place where its habitation shall be placed, as any wise man would. For this purpose it goes among the water-plants, and soon finds a location exactly to its taste. The next thing is to build the framework of its palace, which, spider-like, it makes out of silk cords of its own spinning. These cords are attached to the plants so as to make a room large enough for itself and whatever prey it may happen to capture. As soon as the framework is completed and the rafters are put in their places, it covers the whole with a liquid varnish that makes the walls and roof air- and water-tight, and so elastic as to expand and contract readily. When building it is careful to leave a door open in the under side or floor of the palace, through which it can enter at pleasure. When it is thus far completed it is still unfit for occupancy until the water is excluded.

Fig. 73.—THE WATER-SPIDER AT HOME. (From Wood's *Homes Without Hands.*)

So our little builder sets about the work of filling its new house with air. Now its practical philosophy comes into use. It knows that the air must come from the atmosphere above the water, and must be carried down through the water to the place made ready for it. So it prepares a vessel for this purpose, made of material like that of the palace, and fixes it under its body. (Fig. 73.)

"Carrying its vessel with it, it rises to the top of the water, and giving itself a fling into the air, the vessel is at once filled. It is the air in its clear vessel which gives it the appearance of bright silver or clear pearl when it is seen above the water. Supplied with its load of air, it dives into the water, and, inserting its abdomen in the door of its palace, it empties the air, which, rising to the ceiling, expels some of the water. This operation is repeated till all the water is displaced with air. Now the palace is ready for its occupant, and here it can calmly repose, undisturbed by the commotions which agitate the surface of the pool, and peacefully eat its food and rear its offspring. This picture will give the reader some idea of the air-castles built by this spider. Here you see this interesting little creature in his own home.

"As soon as the mother-spider has finished the work of building, she lays her eggs and encloses them in a saucer-shaped bag which she fixes against the inner side of the cell near the top. In this bag are about one hundred eggs, spherical in shape and very small. This cell, or air-castle, is the spider's true home. Here she makes provision for her offspring, and here she sets her own table, and by her foraging she provides it with all the delicacies of the season. Here the young spider first opens its eyes upon the world to find abundance of food laid up for it by a kind mother. Nothing is required of the large family of infant spiders but to eat the food within their reach and to grow and become strong for the life-duties awaiting them.

" When they are ready to leave home they go forth to employ the same astonishing instinct in the building of similar homes for themselves and their offspring. Thus it has always been. A kind and benevolent God has fitted these little spiders for a life of industry and beneficence, and they enter upon their life-work as though conscious of their destiny, showing a wonderful earnestness and perseverance."

" What an interesting history !" said Charlie. "I think I shall never be driven back from duty because of difficulties that lie in the way. I will always think of this little spider and its air-castle in the water, and never become discouraged. What can be more difficult than to build such a palace surrounded with water ! One globule of air after another is carried down through the water by the persistent spider till the chambers of its palace are filled. It rests not till its work is done. Nothing turns it aside from its great purpose. It means not to live in vain."

" I shall be well paid for my little story if you, dear young friends, will be governed by the lesson

the water-spider teaches you. Like it, I want you
to live with the future ever in view, and to live *for*
the future. Make your deeds and name the inheritance of your race. Live for a purpose. Undertake nothing but that which is great and noble, and
accomplish what you undertake. Be always governed by the true end of human life—the glory of
God—and, like this little spider, you will fulfill the
end of your being."

"We saw another spider," said Henry, rather abruptly—for, boy like, he was becoming a little impatient while Charlie and his uncle were talking
about the moral lessons that spiders teach—"and it
was a very curious spider too—curious in its habits,
I mean. It was seated on a roll of leaves that
looked for all the world as if they had been fastened together somehow just for the spider's own
use, and it was floating on the water. (Fig. 74.)
Every now and then it would leap into the water,
and walk on it as if it were on the ground, and after
a while go back and leap on the raft again. Sometimes it would catch little insects that would happen to fall on the water near it. Tell us about that
spider. What is its name, and how does it make a
living?"

"Among spiders as among men, every spider to
his trade. The spider you speak of would soon die
if it were compelled to earn its living as the water-
nymph does. It makes its living in a very different way, but one much better suited to its tastes

and instinct. Did you notice how much it differed from the other spider in appearance?"

"Yes," said James, "it was larger and much more beautiful. Its color was chocolate-brown, and a broad orange band marked the outline of the upper part of its chest and abdomen. There

Fig. 74.—THE RAFT-SPIDER.

was a double row of small white spots upon the surface of its abdomen, and a number of short dark bars. Its legs were pale red. We all admired it very much."

"It would have been a beauty if it hadn't been a spider," said Mary. "I don't like spiders, anyhow. Tell us what to call it when we see it again. Does it know its own name?"

"I think not, little Inquisitive; at least this spider would need a good memory to remember the long name it has. It is called *Dolomedes fimbriatus.* Here is work for Charlie, with his Latin and Greek lexicons, to find out why such a learned name was given it; but as he has not his dictionaries here, I will try to explain its meaning. The first or generic name is Greek, and means *crafty; fimbriatus* is Latin, and means *fringed;* so we may suppose that the name was given it from its character. When rendered in our language the name is *the fringed crafty spider*, the broad orange band being the fringes of its dress; but because it rides on the surface of the water on a raft it is better known as the *raft-spider*. You observed how it rode on its leaf-vessel, calm as an alderman and as contented as a sailor, as though it had taken the bearings of its rude boat and knew just *why* and *where* it was sailing.

"Its raft is not a chance raft; it designed and built it. It does not know how to spin a beautiful web like that of the garden-spider, but it knows how to make silk threads, and how to use them in putting together the different parts of the raft. It gathers a few leaves and fastens them together with cords, making the raft just large enough to answer its purpose. It then takes its position on it and sails out on the water in search of prey."

" How does it guide its raft?" asked Henry.

" It is borne by the wind or the current of the water. It needs no compass or sails or oars. The surface of the water is alive with insects, which supply our raftsman with game."

" I think it might get sleepy," said Bertha, " sitting so still on its little raft."

" Catch it sleeping! I guess you don't; it didn't build its raft for a sleeping-place. Didn't you see its eyes? How bright they are! And it uses them well, watching keenly for the approach of its prey. I tell you there is a poor chance for a gnat or a May-fly to sport itself in the air if its pupa should be so unfortunate as to come up to the surface of the water near the leaf-boat of our keen-eyed sailor; and if a moth or a fly or a beetle should happen to fall on the water near our hero's boat, its efforts to regain the air would all be vain when once in his clutches.

" But it is not content to sit on its raft and wait till its prey comes within its reach. It is not like some men who sit leisurely in their office, smoking and sleeping by turns, till some chance customer enters and asks for their services. No; it looks out and far off for its customers. When it sees some water-insect at a distance enjoying itself in its quiet sports, it leaves its floating vessel, and, running swiftly over the water, captures its prey and brings it to the raft."

" Good for the spider, but bad for the insect!" said James.

"Insects are not always out of its reach when under water; for sometimes, when it sees a dainty insect which it relishes bathing itself, it crawls down the stems of aquatic plants and catches it beneath the surface. Its ability to breathe for some time under water also protects it from its enemies, for when it sees its enemy coming it hides itself under the raft, and remains there till the danger is over.

"If you look about you among boys and girls of your age—and among men too—you will see not a few who, like the thoughtless insects, are carelessly sporting within the reach of enemies more dangerous than raft-spiders—soul-enemies, that are just ready to devour them. I think of this when I see the little boy-smokers puffing their cigars on the street and hear them using profane or obscene language, imagining that all this is manly. I think of it when I see boys and girls at their weekday sports, or young people riding out for pleasure, on the Lord's Day. Alas! poor immortal insects! the great *raft-spider* is after them, and before they are aware they will be caught in his snares and devoured at his leisure.

"The raft-spider is industrious, frugal and watchful. It spends no time foolishly. It takes advantage of favorable opportunities. It values moments. It never permits delay to deprive it of its victim. So, when you are tempted to put off duty or to waste your time, think of the raft-spider, and be inspired by it to be prompt, industrious, faithful and energetic, and you will become learned, useful and happy."

Just as Uncle Samuel had finished his story the large farm-bell rang out its clear notes, calling all absent members of the family home to supper. Of course our little company did not permit its call to pass unheeded. So, rising at once from their grassy seats by the side of the clear little lake, they returned, Bertha and Mary considering it their special privilege to take hold of their uncle's hands, one on each side, to be led by him.

"I'm glad I'm not a spider," said Mary, "for they have no kind uncles to tell them nice stories."

"Nor any good home to go to, as we have," said Bertha, "for as soon as they are grown up just a little their mothers leave them to make a living for themselves. I'm sure I never could make a living if I had to."

"But God is kind to them," answered Mary, "for he teaches them how to make a living; and they like it too. God knew we could not take care of ourselves till we should grow bigger, so he gave us good parents to care for us and to give us food and clothing."

"Yes, and a good uncle, too, to tell us so many things about God," said Bertha, "and about the poor little insects—and spiders too, for we must remember they are not exactly insects."

Thus they continued recounting God's mercies to them, and expressing their joy that they were so much better off than spiders, and of so much more value, till they arrived at home. Here they were soon engaged in discussing other matters around the supper-table.

We have detailed but a few of the many interesting conversations which Uncle Samuel had with his nephews and nieces during his stay in the country. The limits of this small volume will not permit us to record his stories of the habits and instincts of numerous other insects which were caught by our industrious young naturalists. It would do you good to see the large collection of insects, representing almost all the departments of the insect-world, which they had gathered together by the close of the season, and under the instruction of Uncle Samuel had put in handsome cases. This was the beginning of a cabinet which in after years was greatly enlarged, especially in the department of the Lepidoptera. This cabinet contained not only representatives of home insects, but, as the result of correspondence and exchange, many specimens from other parts of this country and some from foreign lands. Their cabinet is still growing, and their knowledge of insect-life is still increasing, while their collection is an endless source of the purest enjoyment throughout the entire year. Charlie has become quite proficient in the use of the microscope, and is making fresh researches in the different departments of natural science. These researches are deepening his reverence for God and increasing his admiration of the divine wisdom and beneficence as seen in creation.

Bertha returned to her city home greatly improved in health, and with a knowledge of the marvelous facts of insect-life that made her a kind of wonder in the household. She had entered into this study with an enthusiasm that bore her over every difficulty, and greatly quickened her perceptive faculties, so that her improvement was manifest to all. This single summer's communion with one department of God's works under the faithful teaching of her uncle developed her mind more than two years' study in the confined rooms of the city ward-schools could have done; while the attendant physical exercise rendered her vigorous and healthful in body. She did not forget the interesting stories of her uncle, and as she had opportunity she continued to add to her collection of insects till her cabinet became the attraction of all visitors.

If our recital of a few of those interesting talks that Uncle Samuel had with this group of earnest listeners and diligent students of Nature shall inspire the young reader with a taste for the study of the structure and habits of insects, and induce him to reverence the goodness and wisdom of God displayed in this part of the creation, we shall have accomplished the end for which this little book was sent forth.

CHAPTER XIX.

UNCLE SAMUEL has a word to say to the
readers of the "Rambles," and asks the
liberty of adding a chapter for the purpose of telling them how to collect insects, especially butterflies and moths. I very cordially grant him this
favor, for I know how much he loves all young
people, and how greatly he is interested in the
study of the insect department of the great sys-
tem of nature. Here is what Uncle Samuel says:

I hope that some of the enthusiasm that inspired my nephews and nieces in our "Rambles
among the Insects" has reached the readers of the
stories in which these insects have been the chief act-
ors, and that they have a desire to collect specimens
of insects, and by forming a cabinet to prepare them-

selves for the practical study of insect-life. If you will all come into my study and be quiet a little while, I will tell you what kind of an outfit you ought to have, and what you must do with the captured insect in order to prepare it for taking its place in your cabinet, and how you are to place it there. Now, I do not want to talk to you to no purpose; I want to aid you as a band of young inquirers into the instincts and habits of these wonderful little creatures, so that you may know how to gather knowledge and to preserve your specimens for future examination. So you must give me your attention, and when you have put my instructions into practice I want you to write to me and tell me how you have succeeded, what insects you have caught, and how large your cabinet is. Some time in my journeyings I may visit you, and I shall be glad to find you earnest and diligent students of God's works, and to look upon your well-arranged cabinets of Lepidoptera and other insects.

I intend to give you only a few directions which will serve you till you begin to collect an entomological library, and then you will have books that will give you more full and specific instructions about collecting and preserving insects than I can give you in the limited space afforded in one chapter. One of the first books you ought to buy after reading this volume is "Packard's Guide to the Study of Insects," and in it you will find very accurate and full directions to the young collector.

You will also find " Green's Insect-Hunter's Companion " very helpful and suggestive; " Harris's Insects Injurious to Vegetation " is a valuable aid to the young collector. Study these carefully and follow the directions given, and you will soon become expert entomologists. You will find a good many hard words in them, but you can study these out.

One of the first requisites is *enthusiasm.* This is necessary to success in any study; it is specially needed by one who means to study natural objects, not from books merely, but also from Nature herself. On the presumption that you are equipped with this inspiration, you are ready to provide yourself with an

OUTFIT.

As insects are to be caught on the wing, the most important part of the apparatus is the *net.* This can be purchased already made in some of the cities, but collectors must expect to manufacture their own apparatus, and the net can easily be made by attaching to the end of a handle from four to six feet long—an old broom-handle can be made to answer this purpose—a ring from eight to ten inches in diameter, made of brass wire, on which is secured a bag of twenty inches in length, made of mosquito-netting or any thin but strong gauze. Some collectors have a neat cane with a nut at the lower end, into which the net is fastened by a screw. This is a very convenient arrangement, and it utilizes the handle when the collector is walking to the hunting-ground. It should be light enough to be used dexterously with one hand, and when an insect is captured a twist of the net will cover the opening, so that the insect cannot escape. After a little experience you will be able to remove the insect out of the net without damaging its wings or letting it escape. Very small insects ought to be put in your wide-mouthed bottle charged with chloroform, and closed in the bottle, which you can carry with you, a short time, before you attempt to take them them from the net.

If you want to beat trees, bushes, and vegetation for beetles and larvæ, you need a *beating-net.* This is made much stronger, with a shallow cloth bag and having a shorter handle. It is convenient for use in capturing insects that rest on the grass in meadows. The *water-net* may be made like the latter, but the bag should be of grass-cloth or coarse millinet. It is used for capturing insects in the water.

The experience and necessities of the collector will suggest to him various little boxes, vials, and bottles, some containing alcohol or whiskey to be used for beetles and other insects which are preserved in alcohol. For the killing of insects it will be convenient to have a wide-mouthed bottle having an air-tight stopper, inside of which is fastened a piece of sponge. By saturating this sponge with

18

chloroform the bottle will be filled with the vapor. When a moth or other insect is caught let it be put into this bottle, and it will very soon be suffocated. Having brought home from your insect-hunting several fine specimens, you will want to know how to prepare them for the cabinet. The first requisite will be—

SETTING-BOARDS, on which the wings of insects may be spread. These can be made by sawing deep grooves in a thick board of soft wood, and placing a strip of cork or pith at the bottom. If you cannot procure either the cork or pith, you can do without by making a pin-hole in the bottom of the groove for the reception of the pin that holds the insect. The groove should be deep enough to allow a quarter of the length of the pin to project above the insect, and wide enough to receive the body of the insect. The surface of the board should incline slightly toward the groove, as the wings often fall down a little after removal from the board.

The wings should be so set that the hinder margin of the fore wings will be at right angles with the body of the insect. This will enable you to draw the hind wings forward, so as to free their inner margin from the body, and thus expose their forms and markings. When thus arranged they are confined by pieces of soft paper pinned on each side of the wings, so as to hold them to their place. Square pieces of glass, which some use, are liable to remove or injure the scales by their weight. Moths

of medium size and butterflies should remain two
or three days on the setting-board, but the large
moths, such as the Sphingidæ and the Bombycidæ,
should remain a week or more. Dried insects may
be moistened by laying them for twelve or twenty-
four hours in a box containing a layer of wet sand,
covered with one thickness of soft paper. After
they have become moistened and relaxed you can
easily spread their wings.

THE PINS

that are most generally used are the Klager pins.
They are bought in five sizes, so as to suit the largest or smallest insects. These pins are long and
slender, and the insect is set so that one-fourth of
the pin is above the insect. The pin should be inserted through the thorax of moths and similar insects, and through the right wing-cover of beetles.
Many Hemiptera are best pinned through the scu-
tellum.

INSECT CABINET.

If you want to preserve your specimens for permanent exhibition, you will need a cabinet of shallow drawers protected by doors. Such a cabinet
can be so constructed as to form an elegant piece of
furniture, such as will adorn any drawing-room.
The drawer may be eighteen by twenty inches square
and two inches deep, covered with glass. These
boxes should be lined on the bottom with thin slips
of cork. These can be obtained from any cork man-

ufacturer, and are usually twelve by four inches square and one-eighth of an inch thick, costing from $1 to $1.20 per dozen.

If you wish your collection to adorn the walls of your parlor or study, shallow show-cases with a glass front can be made of any size you may desire, but otherwise like the drawers already described. You may then arrange the specimens so as to form a pleasing picturesque group of insects, or, scientifically, under the groups, families, or species. The latter is preferable on many accounts, and the classification is more readily perceived. The insects thus preserved may be kept from insect ravages by securing in the corners of the drawers or show-cases pieces of gum-camphor covered with some porous fabric.

THE REARING OF LARVÆ

is done in a *vivarium*, which every collector should provide for himself. In a small way tumblers may be used, by covering the bottom with moist soil in which the food-plant may be stuck, and drawing gauze over the top. Or small boxes may be used covered with mosquito-netting, but it is better to provide large boxes, according to the wants of the collector, having sides covered with gauze, a glass cover, and a door through which larvæ and their food can be inserted. The bottom should be covered six or eight inches deep with fresh light soil taken from the woods, and this ought to be kept as nearly as possible in its natural condition, so that when the larva penetrates it to pass into its pupal state it will be preserved in a healthy condition. The vivarium should be greatly used by the collector, because insects reared therein are the most perfect, and the habits and instincts of insects can only be readily learned by having them near you for constant observation. The history of new species is thus learned and fresh contributions made to entomological science. I will close this chapter by letting you read a very interesting letter on this subject which was written at my request for your benefit by E. Pilate, M. D., of Dayton, Ohio, an enthusiastic entomologist, and a very successful collector of Lepidoptera :

"The best and surest way to get Lepidoptera eggs is to obtain them direct from *fecundated* females of perfect insects. Chance only may lead to the discovery of eggs deposited on plants or trees, although the female generally lays her eggs *on* or *near* the food-plant of her offspring, and they are concealed under the leaves or in the crevices of the bark. The larva of *Papilio turnus* feeds on apple, tulip-poplar, and wild-cherry trees; *Papilio ajax*, on paw-paw; *Colias philodice*, on clover and pea-vines; *Danais archippus* on *Asclepias syriaca* (common milk-weed); *Limenitis disippus*, on willow; *Limenitis ursula*, on elm, etc.; *Pyrameis cardui*, on thistles; *Junonia lavinia*, on beggar's lice (*Cynoglossum morisoni*); *Pyrameis atalanta*, on nettles and poplar; *Vanessa antiopa*, on willow and poplar; *Grapta interrogationis* on elm and black-berry. Many Sphingidæ live on both wild and cultivated grapevines; Bombycidæ, on oak, willow, and apple, as *Platysamia cecropia; Actias luna*, on walnut, oak, and hickory; *Telea polyphemus*, on oak, sycamore, and rosebush; *Callosamia promethea*, on sassafras and wild-cherry.

" Most larvæ are more or less polyphagous (that is, they will eat several kinds of food), as a matter of necessity more than taste, *when at liberty;* but in captivity very few will eat or even touch any food except that which they feed on in infancy. The search for the chrysalis or pupæ of moths will seldom succeed, except for those in cocoons, which are found in the fall of the year attached to small twigs or under the loose bark of trees, logs, and other hiding-places.

" In warm seasons most chrysalides are hid out of reach at the tops of trees, while the larvæ of the Sphingidæ and Noctuidæ · go into the ground to undergo their pupa metamorphosis. *Citheronia, Eacles, Anisota*, etc. penetrate the soil to a depth of eight to fifteen inches or more."

The best time for digging up pupæ is the latter part of March or early in April, when the frost is entirely out of the ground. Great care must be used, so that the chrysalis is not injured. The larvæ seldom go more than a foot from the tree on which they feed, so in digging it is best to begin about eighteen inches and dig toward the root of the tree.

I now bid adieu to the young readers of the " Rambles," hoping to hear of their great success in collecting and preserving many splendid specimens of insects, and in acquiring an accurate knowledge of their very interesting instincts and of the manners and customs of their domestic life.

THE END.

www.ingramcontent.com/pod-product-compliance
Lightning Source LLC
Chambersburg PA
CBHW021515210326
41599CB00012B/1261